That Savage Gaze
Wolves in the Nineteenth-Century Russian Imagination

Figure 1 Nikolai Yegorovich Sverchkov, *On the Hunt* (1870s). Heritage Image Partnership Ltd/Alamy Stock Photo

The Unknown Nineteenth Century

Series Editor
JOE PESCHIO (University of Wisconsin, Milwaukee)

That Savage Gaze

Wolves in the Nineteenth-Century Russian Imagination

IAN M. HELFANT

Boston
2018

Library of Congress Cataloging-in-Publication Data

Names: Helfant, Ian M., author.

Title: That savage gaze : wolves in the nineteenth-century Russian imagination / Ian M. Helfant.

Description: Boston : Academic Studies Press, 2018. |

Series: The unknown nineteenth century | Includes bibliographical references.

Identifiers: LCCN 2018023234 (print) | LCCN 2018025488 (ebook) | ISBN 9781618118660 (ebook) | ISBN 9781618118431 (hardcover)
Subjects: LCSH: Gray wolf—Russia—History—19th century. | Gray wolf—Control—Social aspects—Russia—History—19th century.

Classification: LCC QL737.C22 (ebook) | LCC QL737.C22 H437 2018 (print) | DDC 599.773—dc23
LC record available at https://lccn.loc.gov/2018023234
ISBN 9781618118660 (ebook) | ISBN 9781644691342

© **Academic Studies Press, 2018.**
ISBN 9781618118431 (hardcover)
ISBN 9781618118660 (ebook)

Book design by Kryon Publishing Services (P) Ltd.
www.kryonpublishing.com

Cover design by Ivan Grave
On the cover: fragments of "The Wolf Hunt" and "On the Road," both by Nikolai Sverchkov.

Academic Studies Press
28 Montfern Avenue
Brighton, MA 02135, USA
press@academicstudiespress.com
www.academicstudiespress.com

For Astrid, Skye and Aidan in our hope that humans will find a way to coexist with wolves and the other animals with whom we share this world

Over the last decade the number of wolves has increased dramatically and continues to do so with each passing year. Wolves are becoming a public pestilence, a national scourge: they run into the cities and even the capitals in broad daylight, and in the villages they approach herds and throttle livestock without fear or danger... Our peasants, who have just recently been freed from slavery, who have scarcely ceased paying a heavy tribute to landowners, have once again fallen into servitude—only this time not to people, but rather to a predatory beast.
—**L. P. Sabaneev,** *The Wolf* (1880)

But I've had a good look, and I've long since been crying. This wolf has been poked through the side with a rake. He breathes through a hole in his side. The air hisses, and it seems as though I hear it through the hole; the edges of his wound move up and down. It's horrible. The wolf's teeth have bit the stick in his mouth; quite close to my face, where it's pressed against the bars, are his eyes. In their corners I see the white part. It's all bloody. His pupils strain straight into my pupils. Unbearable pain, furious hatred, and sorrow are condensed in them, along with a final, hopeless, settled horror.
—**L. D. Zinovieva-Annibal,** *The Tragic Menagerie* (1907)

Table of Contents

A Note on Translation and Transliteration   viii

Acknowledgments   ix

Introduction   xi

Chapter 1   Harnessing the Domestic to Confront the Wild: Borzoi Wolf Hunting and Masculine Aggression in *War and Peace*   1

Chapter 2   The Rise of Hunting Societies, the Professionalization of Wolf Expertise, and the Legal Sanctioning of Predator Control with Guns and Poison   33

Chapter 3   Chekhov's "Hydrophobia," Kuzminskaya's "The Rabid Wolf," and the Fear of Bestial Madness on the Eve of Pasteur's Panacea   70

Chapter 4   Fissures in the Flock: Wolf Hounding, the Humane Society, and the Literary Redemption of a Feared Predator   97

Conclusion   132

Endnotes   140

Bibliography   163

Index   171

A Note on Translation and Transliteration

Russian names and other words in the text, endnotes and bibliography follow a modified Library of Congress system of transliteration without the use of diacritics in which "y" replaces "ii," "aya" replaces "aia," etc. Proper nouns and names are presented in the most common variants found in English-language texts (Alexander rather than Aleksandr, etc.) All translations, unless otherwise indicated, are my own. An imperial *verst* was equivalent to 1.07 kilometers or 3500 feet.

Acknowledgments

There are relatively few scholars studying nineteenth-century Russia through the prisms of ecocriticism and human–animal studies (though that is rapidly changing) and I'm fortunate to have benefited from the advice, friendship and wisdom of most of them. I would like to especially thank Jane Costlow and Amy Nelson, who convened a group of sixteen or so like-minded colleagues at Virginia Tech in 2007 for a conference on animals in Russian history and culture, which was supported by Virginia Tech and Bates College. The group included Cathy Frierson, who had served as my undergraduate thesis advisor twenty years earlier while finishing her PhD at Harvard, and many others whose camaraderie and thoughtful comments on each other's papers made the two-day gathering a paradigm of scholarly collegiality and support. In 2010 an evolved version of my paper, which contained the seeds for this later book project, appeared in the volume that emerged from that conference as "That Savage Gaze: The Contested Portrayal of Wolves in Nineteenth-Century Russian Culture," in J. Costlow and A. Nelson, eds., *Other Animals: Beyond the Human in Russian Culture and History* (University of Pittsburgh Press, 2010), 63-76. Portions of it appear in further incarnation scattered among these pages with permission of the University of Pittsburgh Press.

In the intervening years, as the book slowly unfolded amid the pleasures and burdens of teaching, administration and parenthood I gave a series of papers that would later coalesce into portions of its four chapters at both the national ASEEES conferences and the biennial conference of ASLE. In both venues I frequently co-presented and benefited from the erudition of Tom Hodge and Tom Newlin alongside Jane and Amy, as well as younger colleagues including Molly Brunson (who generously shared her expertise on art copyright permissions), Anya Corke Allen, Isabel Lane, and others. Here at Colgate Nancy Ries, Mieka Erley and Chris Vecsey have read individual chapters and offered sage counsel, as did fellow wolf enthusiast Tovar Cerulli during a campus visit to give an Environmental Studies Brownbag on being a "mindful carnivore." I would like to thank Alice Dautry, formerly Director of the Pasteur Institute in Paris, for her help in obtaining permission to reproduce a photograph of a group of Smolensk peasants attacked by a wolf, who were among the first foreigners to be treated by Pasteur. I am truly grateful to Joe Peschio, editor of The Unknown Nineteenth Century series at Academic Studies Press, former ASP acquiring editors Faith Wilson Stein and Oleh Kotsyuba, and the excellent editing staff at ASP for their encouragement and professionalism. The two anonymous peer reviewers arranged by ASP provided detailed, pointed and astute advice, which I took to heart. The book would not be what it is without their invaluable input.

I benefited from the support of Colgate's Research Council via two major grants, a discretionary grant, a subvention grant, and an Associate Faculty Research Leave in spring 2017. The grants enabled three month-long research trips to Russia in 2006, 2012, and 2014 while the semester of sabbatical allowed me to get the book into final shape for submission. Most of my research for the book took place at the National Library in St. Petersburg, particularly in the rare journals collection, and I am indebted to its hardworking and resourceful staff.

My development as a scholar of ecocriticism and human–animal studies coincided with the transformation of Colgate University's Russian Department into an interdisciplinary program that includes wonderful colleagues across the academic divisions and culminated in my senior joint appointment in Russian & Eurasian Studies and Environmental Studies. Little did I know on arriving here in 1998 that I would be teaching a course called "Hunting, Eating, Vegetarianism" alongside Russian language, literature and culture fifteen years later! My colleagues' and the institution's openness to allowing this scholarly and pedagogical evolution has been instrumental in this book coming to fruition.

Introduction

In 1862, the year after Tsar Alexander II's edict of emancipation freed serfs throughout the Russian empire, a three-page account of an attack on a village by a rabid wolf appeared in the August issue of the Moscow-based *Journal of Hunting*.[1] Entitled "A Horrific Event," the unsigned piece detailed how the wolf had run into the Village of the Evangelists in the Belorussian province of Minsk at midnight on January 27, biting fifty-eight people. The victims included seven women who were nursing and two who were pregnant. The onslaught ended only when a villager attempting to defend his wife stuffed his elbow into the wolf's mouth. This enabled a soldier stationed in the village to kill the crazed animal with an axe. Reinforcing its claims to accuracy, the article then painstakingly enumerated the medical outcomes for those the wolf had attacked, including five additional people it had bitten before entering the village. All were treated with arsenic according to medical protocols of the time under the guidance of a Dr. Grabovsky, but—as he later reported to the Vilnius Medical Society—the treatment was to no avail. The desperate patients also received no benefit from the cures provided by "charlatan" folk healers, who snuck past the practitioners of official medicine to meet with them. In the months following the attack forty-one of the victims died, many of them praying fervently with their families as they awaited the end. Among them were the courageous husband, who died after five weeks, and his wife, who succumbed after three months, orphaning their unweaned child.[2]

This 1862 narrative, while presented as factual and filled with specific detail, stretches a modern reader's credulity. Could one wolf—even if rabid—really attack so many people? Why didn't the soldier or one of the other villagers have a gun at hand?[3] What should we make of the mixture of folkloric, religious, and medical motifs implicit in this dramatic recounting of a beast emerging from the forest at midnight, gripped by a demoniacal frenzy, to terrorize an entire village (and one so aptly named)? How common were such accounts in Russia and what role did they play in both Russians' and foreigners' perceptions of the Russian empire, as well the significance of Russian nature and especially animals in the lives of its peoples?

Such questions form central foci of this book, which explores the powerful presence of wolves in imperial Russian culture. *That Savage Gaze* draws on the burgeoning scholarly fields of ecocriticism and human–animal studies, as well as extensive primary research conducted in Russia, to investigate the ways in which Russians perceived and portrayed the substantial wolf populations that persisted in the Russian empire throughout the pre-revolutionary era. It delves into diverse spheres in which hunters, writers, conservationists, members of animal protection societies, scientists, peasants, government officials and others contested the ecological, economic and cultural significance of wolves. It also elucidates the processes by which the empire's large wolf populations became intertwined with Russian identity both domestically and abroad, revealing Russian insecurities vis-à-vis the countries of Western Europe. Ultimately, it traces the importance of wolves as a symbolic locus for the expression of underlying tensions in imperial society as Russians grappled with the pressures of modernity.

Throughout the nineteenth century and especially in the decades that followed the emancipation of Russia's serfs in 1861, wolves occupied a critical position amid a web of anxieties that preoccupied Russians of all classes for different but intersecting reasons. For a nobility in decline hunting wolves served to shore up an eroding sense of patriarchal entitlement and hegemony. Wolf hunts also allowed gentry hunters to display a distinctively Russian form of masculinity that paralleled their performances in related behavioral arenas such as warfare, gambling and dueling. For members of Russia's intellectual classes including scientists, and its political elites, wolves represented an ongoing challenge as they attempted to impose order and control on the empire's rural landscapes and inhabitants and oversaw efforts to domesticate and utilize its remaining wilderness. Prior to Pasteur's development of a rabies vaccine in the mid-1880s wolves also posed a serious problem for Russia's nascent medical establishment, as ineffectual treatments in the aftermath of attacks by rabid wolves subverted its claims that victims (typically peasants) should place their trust in trained medical professionals rather than seeking the care of village folk healers; this perpetuated distrust of modern medicine in the countryside. For peasants in the empire's most vulnerable regions, wolves represented a perennial threat to their livestock and even themselves. In combatting wolves they revealed superstitions and employed techniques that members of

Russia's upper classes criticized as emblematic of their backwardness and barbarity. For all these reasons, wolves came to serve as a psychic nexus for the abiding anxieties and class tensions that characterized the late imperial period to a degree that was disproportionate to the actual damage they inflicted and dangers they posed.

In tracing these evolving frictions and rifts I focus on nineteenth-century European Russia but situate this focus within a broader geographical and temporal context, as well as amid studies of wolves in other countries and cultures ranging from America to Japan. I enlist the tools of literary analysis alongside those of historical research, explicating canonical texts such as Leo Tolstoy's *War and Peace* alongside neglected or minor literary works while paying equal attention to non-literary sources such as hunting journals, medical tracts, legal codes, natural history treatises, and memoirs. Together, these diverse and wide-ranging sources provide an array of perspectives on the means by which Russians demonized and persecuted wolves, as well as their particular fear of rabid wolves. I juxtapose this widespread antipathy with isolated voices that began to question Russian culture's hostility toward wolves as conservation and animal welfare movements began to coalesce in the late nineteenth century and as Russia's wolf populations decreased. Ultimately, I argue that wolves played a foundational role in Russians' conceptualization of the natural world in ways that reverberated even within the social realm, providing insights into broader aspects of Russian culture and history as well as the opportunities and challenges that modernity posed for the Russian empire.

Russia's Wolf Problem (*volchy vopros*) highlighted the challenges of its efforts to join the "civilized" nations of Western Europe and to overcome both internal and external perceptions that it lagged behind its West European counterparts in social, economic, legal, and political spheres.[4] Throughout the decades that preceded and followed emancipation, wolf populations across the Russian empire remained much higher than those in Western Europe, where wolves had been eradicated or their populations dramatically reduced. Government researchers, the hunting community, and others concerned for the welfare of Russia's newly emancipated peasantry lamented the toll that wolves took on Russia's mostly unarmed rural population and especially children. They also bemoaned the economic damage that stemmed from wolf attacks on livestock. Such attacks were especially prevalent in the more densely populated provinces of European Russia where the majority of peasants engaged in subsistence agriculture and their livestock were concentrated. Rabid wolves caused particular concern given the high incidence of rabies that resulted from wolf bites, the horrific nature of the disease, and the lack of any effective treatment prior to Pasteur's invention of a vaccine in 1885. From the 1850s through the 1890s, stories of wolf packs laying siege to isolated areas of the countryside or of lone rabid wolves attacking many villagers in a single night appeared regularly in the Russian press. Ranging from matter-of-fact reports to highly dramatic accounts like the one above, they contributed to a cultural demonization of wolves that found legal expression in the epochal hunting law of February 3, 1892, which codified Russia's hostility toward wolves alongside other predators.

In keeping with this overriding cultural antipathy, Russian hunters throughout the nineteenth century saw wolves (along with bears) as the ultimate quarry. Early in the century aristocratic hunters on horseback with serf retainers pursued wolves with large packs of dogs including both scent hounds and borzois in a manner that resembled but also fundamentally differed from fox hunting in England. As the Russian landed aristocracy contracted, however, and especially with the abolition of serfdom, hunting on this grand scale gradually gave way to less extravagant and more egalitarian modes of hunting. This accounted in part for the fascination with hunts by members of the imperial family, which maintained these traditions—royal hunting expeditions and kill tallies were described in detail in Russia's hunting journals. In the second half of the century wolves were more commonly pursued by hunting club members drawn from the gentry and newly professional classes in Moscow, St. Petersburg and the provinces; by professional hunters and trappers (whose primary interest lay in other more marketable and less challenging quarry); and in some cases by peasants defending their livestock, as they had the right to hunt on land belonging to their communities (although they often leased this right to individual members of the gentry or hunting clubs). Amid these changes—which epitomized the close linkage between the Russian empire's changing social order and the population's interactions with nature—gentry and other privileged hunters continued to assert that their self-appointed role as the protectors of rural peasants necessitated their killing wolves, just as they continued to believe that wolves provided some of the most challenging and desirable hunting experiences. The members of hunting clubs often paid supplemental fees in addition to their yearly membership dues in order to participate in winter wolf hunts organized by Russia's hunting societies as part of wolf reduction campaigns. This mixed set of motivations led some to oppose as "unsporting" a new means of killing wolves that was controversial not only because it differed so radically from what had come before but also because it seemed to threaten the very notion of what wolf hunting represented: poisoning wolves with strychnine. Pioneered in wolf eradication campaigns in Western Europe and North America and promoted by the tsarist government and prominent wolf experts, poisoning became a point of particular controversy in heated debates over Russia's wolves that emerged in the 1870s.

In the late 1880s and 1890s, as the debate over the Wolf Problem continued to seethe isolated voices began to question Russian culture's unwavering antagonism toward wolves and even the ethics of hunting itself. Most prominently Tolstoy, who had hunted avidly until his spiritual crisis of the late 1870s and early 1880s and who portrayed an aristocratic wolf hunt so memorably in twenty mesmerizing pages of *War and Peace*, renounced hunting and became a pacifist and vegetarian. In a preface to an 1890 article entitled "A Wicked Pastime" by his friend and collaborator Vladimir Chertkov, which appeared in the newspaper *Novoe vremia* (*A New Age*), Tolstoy called for others to renounce hunting as well. The article hinged on Chertkov's own renunciation of hunting after an epiphany—spurred by his detailed recollection of beating a wolf he had shot with a heavy stick on the bridge of its nose until it died as it stared at him with savage rage (*s dikim isstupleniem*)—that his lack of empathy for the wolf undermined his own humanity.[5]

The Imperial Russian Society for the Protection of Animals, established in 1865, also ran a series of articles and stories around the turn of the century that portrayed wolves in a more sympathetic manner. These polemicized with the anti-wolf views that continued to prevail among hunters and the public. They criticized especially fervently such institutions as wolf-hounding competitions, in which rival packs of borzoi hounds pursued, mauled, and often killed wolves that had been previously captured and were released from their cages in front of spectators for these events. These spectacles attempted to recreate the borzoi hunts that had predominated earlier in the century but did so artificially and amid a changed social and cultural landscape in which they came under fierce criticism as a symbol of Russian backwardness and cruelty by members of Russia's emerging civic society, including influential writers like Anton Chekhov. The rise of natural history, with its attempts to analyze Russia's flora and fauna in scientific terms, also helped to reduce longstanding folkloric associations of wolves (and especially rabid wolves) with werewolves and even vampires. Pasteur's development of a rabies vaccine and the rapid establishment of rabies stations across the Russian empire over the next decade, which coincided with reduced wolf populations in European Russia due to wolf eradication efforts, also helped somewhat to alleviate the terror that wolves inspired in many rural Russians. Despite these cultural fissures, however, antipathy toward wolves and fear of rabies continue to hold sway in Russian culture throughout the tsarist era and into the post-revolutionary period. The vast majority of Russians continued to view wolves as a scourge to be persecuted and, if possible, eliminated.

Figure 2 Bloodthirsty Wolves Attacking Travelers. Popular print (1894). Image copyright Grafika

Grappling with the rich complexity of these intertwined historical, cultural, ecological and otherwise multi-dimensional issues represents a daunting task. How, for example, should one interrelate and disentangle attitudes toward wolves among a rural population that consisted mostly of peasants engaged in subsistence agriculture (who possessed detailed local knowledge but often were also influenced by folklore or superstition) with those of gentry hunters and members of Russia's emerging scientific establishment (who drew on a wider set of more "authoritative" perspectives, including those of writers and scientists abroad)? How did Russia's early medical professionals relate to folk medicine in the treatment of rabies prior to and during the shift to Pasteur's method of vaccination, which quickly supplanted all other approaches, and how did uneducated commoners view both the official medical establishment and local healers? How did attitudes toward wilderness change during the nineteenth and early twentieth centuries as European Russia became increasingly "domesticated" and wildlife populations diminished, and what role did this play in changing attitudes toward wolves? Did Russians relate to these various issues differently than their Western European and other counterparts abroad including American settlers, who also confronted large populations of wolves?

In meeting these methodological and conceptual challenges I will draw on a wide range of scholarship including the emerging fields of ecocriticism and animal studies, which have recently begun to make inroads into Slavic studies, as well as works of environmental history. While this book focuses primarily on cultural history, I will also refer selectively to the ongoing and robust discussion of wolves in the fields of wildlife and conservation biology. It's worth noting at the outset that there has been relatively little cross-fertilization between Western and Russian scholarship on wolves, particularly in fields other than zoology. One of my major goals, in fact, is to address this lacuna by employing current Western scholarly approaches in my analysis of the historical significance of wolves in Imperial Russia.

One may view both ecocriticism and animal studies as having developed out of cultural studies with animal studies representing a branch of ecocriticism.[6] As is true of cultural studies in general, ecocriticism attempts to achieve broad understandings of its subjects by ranging widely through such diverse disciplines as history, anthropology, literature, journalism, the visual arts, and the natural sciences. The fundamental question that ecocriticism explores is how humanity views itself in relation to nature and how these views vary between societies and over time, particularly as expressed in literature and the arts. Animal studies, at its best, shares this breadth of perspective with particular attention to the ethical and ecological implications of humanity's interrelations with other animals; hence many in the field, myself included, prefer the more encompassing term human–animal studies. A unifying theme of human–animal studies is that the ways in which humans view and portray other animals lend important insights not only into our interactions with and influences on the natural world but also into our conceptions of self and society.[7]

During the last decade Slavic studies has begun to incorporate perspectives from these increasingly significant and often intertwined fields alongside works in environmental history more broadly. For example, the preeminent journal *Slavic Review* devoted its spring 2009 issue to the theme of "Nature, Culture, and Power." *Other Animals: Beyond the Human in Russian Culture and*

History was published a year later. This collection of multidisciplinary forays into the significance of animals in Russian culture includes my first publication on Russia's wolves alongside treatments of bears, dogs, folk veterinary medicine, reindeer herding, literary representations of animals from the nineteenth century to the present, and other topics. In their introduction to the collection, editors Jane Costlow and Amy Nelson emphasize the persistence of Western tendencies to see "Russia as the wild other of Europe, an exotic (or pathetic) human creature who is somehow closer to nature and the animal realm."[8] I will show that wolves played a significant role in Western perceptions of Russia in ways that echo and reinforce this statement. Costlow's *Heart-Pine Russia: Walking and Writing the Nineteenth-Century Forest*, published in 2013, utilizes ecocritical perspectives to explore the importance of forested landscapes in Russian cultural history; it ranges widely through scientific writing and forestry journals alongside the works of writers and visual artists in a manner that has helped to guide my own approach.[9] Henrietta Mondry's *Political Animals: Representing Dogs in Modern Russian Culture*, published in 2015, emphasizes the often transgressive cultural significance of dogs from the eighteenth century onward in a range of contexts and a variety of literary and non-literary genres and mediums.[10] Most recently *Canadian Slavonic Papers*, noting the "explosion of interest in the various ways that animals have shaped human identities and experiences," issued a call in 2016 for submissions to a special issue on "Animals in Eastern Europe and Russia." Slavic studies has clearly begun to embrace these fecund approaches.[11]

Surprisingly, no previous scholars have delved with any persistence into the significance of wolves in imperial Russia utilizing these perspectives, whereas a plethora of scholarly and popular books have addressed the ecological and historical significance of wolves across a variety of other cultures. Most Soviet era scholarship on wolves was zoological and pragmatic in orientation, and contemporary Russian approaches do not tend to draw on current scholarship in ecocritical and human–animal studies, which has largely been driven by Western scholars. Moreover, Western studies—even those that strive for comparative and international perspectives on wolves—have generally included minimal information about wolves in Russia, reflecting linguistic challenges and a less robust tradition of scholarly collaboration between Russian experts and those beyond the Soviet bloc and now post-Soviet space. For example, a substantial 2002 collaboration of eighteen mostly European researchers that aimed at systematically documenting and evaluating historical accounts of wolf attacks on humans across the globe bemoaned at the outset the group's inability to enlist a Russian co-author.[12] An authoritative 2003 compilation entitled *Wolves: Behavior, Ecology, and Conservation* co-edited by the dean of American wolf studies, L. David Mech, and Italian wolf expert Luigi Boitani, devoted relatively little attention to Russia and Eurasia.[13] A recent densely researched monograph on wolves by British human–animal studies scholar Garry Marvin utilizes my 2010 chapter in *Other Animals* as its primary source for information about wolves in Russia.[14] A 2015 collection entitled *A Fairytale in Question: Historical Interactions between Humans and Wolves*, which includes fourteen mostly collaborative chapters on wolves throughout Western Europe and North America, alludes to the history of wolves in the Russian empire only in passing and includes just one chapter on the post-Soviet space—a discussion of recent wolf attacks in Central Asia.[15]

Among Russian experts, the culmination of Soviet-era understanding of wolves appeared in 1985 in a 600-page volume with more than twenty contributors edited by the prominent zoologist D. I. Bibikov and published by the USSR Academy of Sciences entitled *The Wolf: History, Systematics, Morphology, Ecology*.[16] Although primarily zoological in orientation, the volume included a brief discussion of human attitudes toward wolves, as well as wolf predation on both livestock and wildlife, and—while it focused primarily on wolves within the Soviet Union—acknowledged the work of key international wolf scholars. Its importance was underlined by its translation into German two years later.[17] In addition to Bibikov, the other Russian wolf expert who is most often referred to in Western scholarship is M. P. Pavlov, whose 350-page monograph *The Wolf* appeared in 1982 (and in its second edition in 1990).[18] Pavlov has drawn particular attention for his emphasis on attacks by non-rabid wolves on people and especially children, which have been both cited as evidence and critiqued as exaggerated by subsequent scholars.[19]

The most extensive treatment of Russia's wolves by a non-Russian that I've encountered is an idiosyncratic book containing a hodgepodge of material compiled by a retired US government employee whose explicit goal is to discredit the sentimental views of wolves that he believes characterize Western environmentalism. Will N. Graves' *Wolves in Russia* gathers a substantial variety of information related primarily to Russia's wolves in the twentieth century. Unfortunately, it is a tendentious work that spends only a few pages on the pre-revolutionary period and does not meet scholarly standards of rigor, organization, reliable and consistent citation, or theoretical sophistication.[20] In addition, a German book published in 2011 provides the full Russian text of an important government brochure of 1876 that compiled statistics about Russia's wolf populations and the damage they caused to rural agriculture along with other accounts by contemporaries and some briefer writings on wolves by present-day contributors. Taken as a whole, the volume promotes an intensely negative portrayal of wolves, drawing on the Russian historical experience to support this stance.[21]

In contrast to these examples, two relatively recent works of environmental history provide excellent paradigms of how one can approach the interwoven stories of wolves and humans across centuries in two very different cultures: colonial America and pre-modern Japan. In *Vicious: Wolves and Men in America*, historian Jon Coleman traces the contours of human persecution of wolves in colonial America, touching on recent reintroduction efforts briefly in the book's final pages. His treatment draws together history, folklore, the colonists' Christian beliefs, and wildlife biology to explain their strikingly savage treatment of wolves, who threatened not only their livestock but their dream of domesticating the wild landscapes of the New World. One of Coleman's especially instructive insights is that the tales colonists told each other about wolf attacks on defenseless humans inversely mirrored reality, in which it was humans who killed wolves.[22] Indeed, one quickly realizes that the "vicious" of his title refers to human attitudes and actions over time toward wolves, rather than describing the wolves themselves.

Historian Brett Walker explores the history of wolf veneration and subsequent extermination in a very different cultural context in *The Lost Wolves of Japan*. Like Coleman, but with particular attention to the place of wolves in folk culture and the Shinto religion, Walker traces the shift from traditional Ainu beliefs that wolves deserved veneration, to rising fear of rabid wolves in the eighteenth century, to persecution of wolves through hunting and poisoning in the nineteenth century that culminated in their extinction in Japan by 1905. He describes how modernity, including scientific progress and the rise of ecological thinking in the twentieth century, caused fundamental shifts in Japanese cultural attitudes toward wolves, anchoring his analysis in a multidisciplinary and comparative context.[23]

These exemplary works of environmental history illustrate some of the ways in which varying climates, landscapes, flora, and fauna can interact with religious, economic, and cultural differences to become interlinked with humans' senses of themselves and the roles that animals have played in their histories. My approach emulates such works but also reflects my background as a literary scholar with expertise in cultural studies, ecocriticism, and human–animal studies rather than as a historian. I will draw consistently on literature as a source of insight into Russian attitudes toward wolves throughout the book, and my readings of memoiristic and journalistic accounts will also pay close attention to their narrative qualities. At the same time—as in my earlier work on gambling as a cultural institution and central component of gentry identity in nineteenth-century Russia—I will range widely through a variety of non-literary sources in order to grapple with the fundamental questions at hand.[24] While *That Savage Gaze* does provide a history of imperial Russians' attitudes toward and treatment of wolves during the nineteenth and early twentieth centuries, and does so in roughly chronological order, it does this through focusing on certain key tropes, texts, and contested areas of cultural discourse rather than by attempting to exhaustively catalogue every instance in which one encounters wolves in nineteenth-century Russian sources. Nor does it provide a comprehensive overview of wolf populations and distribution in the imperial period. This reflects both the problematic and conflicting nature of much of the information available on wolves in imperial Russia and the fact that zoology was only beginning to establish itself as a scientific discipline in this period, as I will explore in Chapter 2.

The primary sources that populate the book derive from three month-long research trips to St. Petersburg in 2006, 2012, and 2014. During these visits I delved into a variety of neglected nineteenth-century texts at the National Library in St. Petersburg, focusing increasingly on wolves. Among the most significant of the resources I explored is a trove of imperial hunting journals that have received little attention even among Russian scholars. As one of the major venues in which Russian attitudes toward wolves were on display, they comprise a key source of the primary materials utilized in the immersive analysis of Russian culture for which the book aims. Most are unobtainable outside Russia but available as full hardbound runs in the National Library's journal collection.

Reflecting Russia's "thick journal" tradition in their eclectic nature, imperial Russia's hunting journals employed a variety of genres ranging from the literary to the scientific in attempting to address issues of interest to hunters and the general public. They contained hunting stories by authors famous and unknown, poems about nature, letters from provincial readers reporting on local hunting conditions and animal populations, hunters' travel journals from trips abroad, official communications from regional hunting clubs, recurrent examples of such genres as "a hunter's memoirs," scientific articles in the disciplines of zoology, geography, and ethnography, and early examples of conservationist literature. These diverse genres coexisted through their common emphasis on issues of concern to hunters and others interested in Russia's flora and fauna. The journals provided a forum through which editors, contributors, and readers of various backgrounds shared a wide range of perspectives and knowledge. While imperial Russia's hunting journals serve as a particularly apt example, many of the other materials that inform this study have been similarly neglected. These include memoiristic and overlooked literary accounts of wolves and wolf hunting, Russian hunting laws and commentary on legal codes, nineteenth-century newspapers, and forestry, humane society, and medical journals. Taken as a whole, this range of sources provides the possibility for a multifaceted description of evolving Russian attitudes toward wolves during the nineteenth and early twentieth centuries.

One of the primary challenges in writing about animals is that they do not have readily intelligible voices and do not themselves leave documents, as Erica Fudge has highlighted: "The only documents available to the historian in any field are documents written, or spoken, by humans."[25] Therefore, any history of animals is in fact a "history of human attitudes toward animals."[26] This renders problems of subjectivity and perspective especially proximate, which perhaps helps to explain a general tendency among practitioners of ecocriticism and human–animal studies to acknowledge and sometimes even welcome the subjective aspects of our scholarly endeavors.[27] Fudge divides animal histories into three types: intellectual, humane, and holistic. In the first, animals represent a means for better understanding major intellectual currents of the era under study, as exemplified in books that trace the use of animal imagery to illuminate "the medieval mind."[28] Her second category focuses on "the lived relation" between humans and animals, wherein what humans say and write about animals reveals something about themselves as well. Hilda Kean's *Animal Rights* (1998) exemplifies this approach in her view: it investigates animal welfare movements in nineteenth and twentieth-century Britain but interrelates them with changing social relationships, including the rising independence and influence of women. Fudge's third category of holistic histories goes even further in utilizing the representation of animals to recast our understanding of a culture's past. Kathleen Kete's *The Beast in the Boudoir* (1994), which draws on the literature of petkeeping in nineteenth-century Paris not only to explore the relationship between humans and their pets but also to better understand the rise of bourgeois ideology and changing attitudes to class and gender, represents a preeminent example of this type of history, as Fudge explains.[29]

My study, like those of Coleman and Walker, attempts to trace how Russians' evolving attitudes toward wolves shed light on a society in a state of flux. In this sense it straddles the second and third categories of Fudge's tripartite categorization, the borders of which she emphasizes are permeable. The roughly half-century between the emancipation of 1861 and the revolutions of 1917 represented an extraordinarily difficult time for imperial society and the tsarist regime and exerted particularly excruciating pressures on the peasantry. Tracing this turmoil through the prism of the treatment of wolves renders crucial aspects of the period in sharp relief. Many of the scenes I will explore involve members of the gentry and educated elite such as doctors, scientists and writers interacting with serfs, peasants and other commoners against a backdrop of Russian nature and its most feared predator. In some cases these encounters yielded a sense of kinship rooted in common perceptions of a shared enemy. In others the protagonists' attitudes diverged, revealing deep fissures of understanding as all segments of Russian society attempted to navigate the empire's changing social, economic and environmental landscapes.

If communication across social and class divides was fraught and problematic, then attempts to peer across the divide between species was infinitely more so. Throughout this book I will investigate human attempts to interpret wolves and their behaviors. Hunters attempted to understand wolves in order to more effectively kill them. Russia's early zoologists strove to ascertain wolves' and other predators' habitat and prey requirements, breeding and hunting activities, and social relationships in order to create scientific knowledge and inform government policies. Doctors endeavored to gauge the potency of the rabies "poison" carried by wolves and understand the symptoms and progression of "canine madness" in order to help their patients. Peasants, who often lived in the closest proximity to wolves, sometimes viewed them through the lens of superstition but also could be among their most astute observers. Writers of fiction for both children and adults attempted to capture wolves' inner worlds with varying degrees of realism, sentimentality, and anthropomorphism.

Many of these instances will culminate in moments at which the human gaze intercepts or interlocks with that of the wolf, which meets or avoids it. The human observer will interpret the wolf's gaze variously as savage, frenzied, demonic, resigned, despairing, inscrutable, alien, or (rarely) kindred. This recurrent trope will serve as a touchstone for tracing evolving attitudes toward wolves and larger questions of humanness, wildness, otherness, and the complexities of the human–animal divide. My study straddles the era during which John Berger argued in his famous 1977 essay "Why Look at Animals?" that industrialization and urbanization caused much of humanity to lose its direct contact with animals other than the artificial intimacy we enjoy with pets or the intimate alienation we experience in observing animals in zoos. In the wake of modernity, in Berger's view, humans and animals view each other "across an abyss of non-comprehension... [and we humans] across ignorance and fear."[30] The fascination and challenge of responding to and interpreting the gaze of animals has become a point of pressing concern for modern human–animal studies, as summarized by Philip Armstrong in a 2011 book chapter "The Gaze of Animals." Armstrong traces the ways in which human interpretations of the animal gaze have shifted dramatically over the centuries, correlating to larger epistemological shifts ranging from the rise of modern science to changing literary forms.[31] I will highlight moments involving the human-wolf gaze and interrelated ones such as the practice of imitating the howls of wolves to elicit a response as we encounter them throughout the book, then consider this issue more fully in the conclusion in the hopes of contributing to this ongoing discourse.

Chapter 1 of *That Savage Gaze* focuses on Tolstoy's protracted portrayal of a wolf hunt in *War and Peace*, fictionally set in 1810, which encapsulates key features of the grand hunt on horseback with hounds and borzois (*psovaya okhota*) that provided a behavioral platform for a specifically Russian sort of aggressive masculinity. This aristocratic social institution, which rested on the crumbling foundation of serfdom, would be supplanted by other methods of hunting wolves during the course of the century, but it remained as a key cultural backdrop for these later approaches. The chapter situates Tolstoy's fictional representation in this larger socio-historical context by drawing on sources ranging from memoirs, to minor literary works, to articles in both the popular press and Russia's hunting journals. It pays particular attention to a work published in 1859 by the little-known writer E. E. Driansky entitled *Notes of a Hunter with a Small Leash of Hounds*. Driansky's pseudo-memoir, along with other contemporary sources, provides a window into the lexicon and cultural practices that underlay Tolstoy's more famous representation of aristocratic wolf hunting. The chapter also briefly compares the Russian wolf hunt with its British counterpart, the fox hunt.

Chapter 2 juxtaposes imperial Russia's two major hunting societies and their roles in controlling the empire's wolf populations in the context of the emerging science of natural history. The Moscow Hunting Society, established in 1862, and the Imperial Society for the Promotion of Game and Wildlife of Economic Significance and Proper Hunting, established in 1872, addressed Russia's Wolf Problem and strove to reduce wolf numbers through divergent approaches. The former leased land accessible to its exclusive Moscow membership and arranged coveted gun hunts for wolves during which members benefited from the skills of wolf experts retained by the Society. The latter attempted to analyze wolf populations throughout the empire and encouraged poisoning of wolves alongside hunting. It also published imperial Russia's most significant monthly hunting journal, *Nature and Hunting*, and a weekly newspaper, *Hunting News*. The chapter details the work of the two successive editors of these publications, L. P. Sabaneev and N. V. Turkin. Sabaneev, trained as a zoologist, published a lengthy monograph on wolves in 1880. Turkin developed into the empire's foremost expert on Russian and international hunting law and served as a leading architect of the 1892 hunting law. Together, their contributions were central to institutionalizing wolf reduction efforts in imperial Russia.

Chapter 3 explores the significance of rabid wolves and rabies in the Russian empire during the pre-Pasteur era, as well as the profound impact of Pasteur's 1885 vaccine. It situates neglected literary portrayals of attacks by rabid wolves in Chekhov's "Hydrophobia" and T. A. Kuzminskaya's "The Rabid Wolf" (both published in 1886) within the framework of journalistic accounts about Russia's Wolf Problem and medical texts on rabies from 1780 through the 1880s. It also utilizes perspectives drawn from Charles Rosenberg's influential 1992 essay "Framing Disease." Together, these literary and non-literary sources provide a window into the anxieties of the post-emancipation era as rural Russians of all classes, but especially the emancipated peasantry, found themselves painfully positioned between the tracts of untamed nature that lingered in Russia—symbolized by the potent figure of the rabid wolf—and the forces of modernization and social change imposed by provincial authorities and exemplified in the newly institutionalized forms of medicine that coexisted uneasily with traditional folk healing. The chapter ends by describing the journey of nineteen Smolensk peasants who'd been bitten by a rabid wolf to Paris for treatment by Pasteur in early 1886, as well as the subsequent establishment of rabies stations throughout the Russian empire.

Chapter 4 complements portrayals of wolves in earlier chapters by highlighting shifting attitudes toward the turn of the century as certain voices began to question Russian culture's demonization of wolves and other predators. It emphasizes the significance of the Russian Society for the Protection of Animals (RSPA), established in 1865, which vehemently opposed such practices as wolf hounding and poisoning and was especially concerned about the relationship between animal abuse and the plight of Russia's rural poor. It explores the importance of literary and memoiristic works in promoting the glimmerings of a more compassionate sensibility toward wolves that emerged among some Russians toward the turn of the century. These range from Chekhov's 1895 story about a nurturing mother wolf, "Whitebrow," to a series of stories and articles that appeared in the RSPA's monthly journal and elsewhere, some of which were narrated from the perspectives of wolves, to a 1907 first-person narrative in which a young girl encounters a wolf wounded in the "tsar's hunt." The chapter probes the difficulties that writers faced in trying to represent wolves' perspectives, including the attractions and perils of portraying animal protagonists anthropomorphically.

The conclusion reprises the unique role that wolves played in imperial Russian history. It revisits and muses on the trope of the interpenetrating gazes of humans and wolves with reference to modern human–animal studies. It addresses the related issues of subjectivity and selectivity that inevitably underlie a project such as this one. Finally, it briefly addresses how our attitudes toward wolves in the present-day can be informed by the Russian experience.

CHAPTER 1

Harnessing the Domestic to Confront the Wild: Borzoi Wolf Hunting and Masculine Aggression in *War and Peace*

> There are still vigorous Russian hunters in whose veins flows not French bouillon or German savory soup but pure Russian blood with that same valiant daring of old, which thinks nothing of "taking down an old wolf by the blade" or "binding alive" the troublemaker. These people don't live in the capitals but in the provincial backwoods, in the forested silence, and the limitless steppes.[32]
>
> —P. M. Machevarianov,
> *Notes of a Borzoi Hunter of Simbirsk Province* (1876)

Tolstoy's Count Nikolai Rostov, an aristocratic protagonist in the vast historical novel *War and Peace* (1865-69), watches ecstatically as his favorite borzoi throttles an old wolf that his pack of more than 100 dogs has pinned to the ground:

> That moment, when Nikolai saw the dogs swarming over the wolf in the ditch, saw under them the wolf's gray fur, his outstretched hind leg, and his frightened and gasping head with its ears laid back (Karai had him by the throat)—the moment when Nikolai saw that was the happiest moment of his life.[33]

2 | That Savage Gaze

Rostov's delight in his borzoi's prowess and lack of compassion for the struggling wolf are conjoined elements in a tableau of domination over nature that culminates Tolstoy's extended portrayal of a borzoi wolf hunt, which represents one of his most protracted descriptions of human interactions with nature in the novel. The joy that Rostov takes in his vicarious victory over the wolf during this "happiest" moment of his life reflects the deep and symbolic significance that wolf hunting assumed for its devotees in Russian gentry culture. Wolf hunting with borzois provided Russian hunters with a visceral yet mediated opportunity to confront a fierce predator that embodied the threat and challenge of Russian's remaining wilderness. The most daring hunters would even grapple with the wolves themselves alongside their dogs in order to bind or stab them. Such displays of masculine prowess had a uniquely Russian inflection, as by the nineteenth century wolves had been eradicated or decimated throughout most of Western Europe but were still quite common in European Russia and elsewhere in the Russian Empire.[34] As the century proceeded, however, this cultural institution gradually waned in the face of the tumultuous changes that accompanied the emancipation of Russia's serfs in 1861. By the time the tsarist regime was engulfed in the revolutions of 1917, hunting wolves with borzois was largely a forgotten relic of past eras.

An 1842 article in the second issue of what would become one of pre-revolutionary Russia's longest-lived hunting journals—*The Journal of Horse Husbandry and Hunting*—contained a fifteen-page rendition of a wolf hunt that took place at the end of the eighteenth century. This was a decade or so prior to Rostov's fictional encounter with a wolf, written in the late 1860s but fictionally set in 1810, which will serve as the centerpiece of this chapter. Entitled "Beast: The Borzoi Wolfhound of Brigadier General Prince G. F. Boriatinsky," the narrative explicitly styled itself on the English tradition of extolling the exploits of noteworthy sportsmen and their dogs, noting that Russia's rich hunting traditions had not received the attention in print that they deserved.[35] The author, Memnon Volunin, asserted that multiple eyewitnesses had corroborated his historical account of the "speed and tenacity" (*rezvost' i tsepkost'*) of the General's eponymous borzoi, "Beast," and the bravery of his master in their encounter with a marauding wolf. His favorite version, and the one on which he claimed to base his retelling, was related to him by Prince Boriatinsky's 85-year-old wife after the Prince's death.

Volunin's narrative encapsulates key features that underlay many nineteenth-century accounts of borzoi wolf hunts, both historical and fictional, up to and including Tolstoy's portrayal in his novel. Like Tolstoy's protagonist

Count Nikolai Rostov after his marriage to Princess Marya Bolkonskaya at the end of *War and Peace,* Prince Boriatinsky is a former military commander who is both patriarchal and occasionally irascible in his dealings with his 2,500 serfs. He sets out with a party of men on horseback and his hunting dogs after receiving reports from nearby peasants that wolves have attacked and killed some of their suckling pigs. The princess—whose perspective the narrative mirrors right down to the use of first-person feminine verb forms—waits anxiously at home. Eventually, a servant dispatched from the hunting party arrives to deliver the reassuring news that her husband, after he and Beast outpaced the rest of the hunting party, has single-handedly captured the wolf by jumping down on it from his horse and clinging to its ears while Beast gripped its throat until his nephews rode up to aid him. The story culminates in the general's triumphant return, accompanied by the captured and bound wolf, to his wife's tremulous welcome: "beside him was stretched out a wolf so terrifying that I have not seen one more fearsome in my life: a forehead like a bull's, and eyes utterly red."[36] The narrator then laments the dissolution of this idealized patriarchal order in which the prince rode out to protect his frightened serfs and family from the fearsome predator. He regretfully notes that the balcony on which the princess awaited her husband's return now lies in ruins.

Volunin presents each of his key dramatis personae—aristocratic hunter on horseback, borzoi, and wolf—in conformity with a mythologized version of the cultural institution of borzoi wolf hunting in Russia. Brigadier General Boriatinsky embodies a protective patriarchy as he physically subdues the wolf on behalf of his peasants and family with the aid of his faithful yet ferocious borzoi, Beast. His borzoi is a liminal creature, which bridges the gap between the domesticity that Boriatinsky is protecting and the wolf, which emblematizes the savagery of undomesticated nature.[37] The borzoi's name—*Zver'* ("Beast")—captures this liminality in all its contradictions, as wolves themselves were often called *zveri* in this era. Beast shares the wolf's ferocity but serves his master in subduing the fierce predator. Significantly, the prince brings the bound and trussed wolf home alive—possible in the aftermath of a hunt with dogs but not usually when gun hunting—so that his wife and household are able to witness his dominance and its captive submission, encapsulated in the rage she attributes to its "utterly red" eyes.

In this chapter I will explore Tolstoy's extensive portrayal of borzoi wolf hunting in *War and Peace* and embed it in the largely forgotten socio-historical context that is presented in such archetypal form in Volunin's account. Borzoi hunting had its heyday in the late eighteenth and early nineteenth centuries but persisted through the waning decades of the nineteenth century until the social, environmental, and political changes that presaged and followed the emancipation of Russia's serfs brought about its inevitable decline. Serf labor had provided the wealth that allowed for the large kennels of twenty to forty scent hounds and a dozen or more borzois that were typical of the largest kennels, at the apex of which was the imperial hunting establishment. In addition, the hunters and handlers who took care of the scent hounds and borzois were typically drawn from among a landowner's serfs. It is no accident that the chapters on the hunt in *War and Peace* immediately follow one in which Tolstoy describes the Rostov family's financial difficulties—the elderly count has mismanaged his own and his wife's property, in part through extravagances such as maintaining a large kennel of dogs—and Nikolai's ineffective attempts to grapple with them by confronting the family's steward. Nor is it accidental that one of their non-aristocratic hunting companions complains that his borzoi cannot be expected to compete against the dogs of the Rostovs and their wealthy neighbors the Ilagins: "you paid a village for each of them, they cost thousands," he conjectures.[38] It turns out that he is exaggerating, but in fact Ilagin "had given his neighbor three families of house serfs a year before" for his red-spotted borzoi bitch, Yerza.[39] The hunt is thus directly linked with the Rostov family's aristocratic mores and means, which rest precariously on the crumbling bedrock of the institution of serfdom that supports their extravagant lifestyle.

Borzoi experts and hunters lamented borzoi hunting's decline and documented this cultural institution in ways that provide a detailed context for Tolstoy's depiction. The sources that I will draw on to elucidate this sociohistorical context range from an extended memoir of borzoi hunting that appeared in 1859, to three book-length guides for the breeding and use of borzoi and scent hounds published in 1846, 1876, and 1891, to briefer articles from the popular and hunting press of the day. Together they will help us to understand how Tolstoy's portrayal of the Rostov family, their huntsman Danilo, and the other gentry hunters they encounter—as well as the domestic animals who mediate their relationships with the prey they seek and especially their ultimate quarry of a mature wolf—reflects the semiotic codes of borzoi wolf hunting as cultural practice in the Napoleonic era. In addition, this approach will help to elucidate the link between borzoi wolf hunting and "Russianness" in part through explicating the lexicon and implicit presumptions that informed Tolstoy's depiction as well as the role that wolves played in Russians' perceptions of their countryside and Russian nature. While the

English foxhunt—most recently explored by animal studies scholar Garry Marvin—and the Russian borzoi wolf hunt had many similarities, they were also strikingly different.⁴⁰ They reflected different cultural attitudes and the much more significant tracts of wild nature that remained in European Russia, not to speak of Central Asia and Siberia, which Russians rightly differentiated from the more domesticated and groomed landscapes of Western Europe and especially England.⁴¹ Wolves, eliminated or reduced to remnants of their former populations in nineteenth-century Western Europe, embodied and symbolized Russia's untamed nature and became interlinked with Russians' views of themselves, sometimes with positive and sometimes with negative connotations. A vast gulf separates the mythic and viral masculinity of the Russian borzoi hunter Boriatinsky, leaping bodily down from his horse onto the back of a wolf alongside his borzoi, Beast, from the aristocratic English foxhunter watching from a dignified distance as a pack of foxhounds mauls and dismembers a cornered fox without the riders' direct intervention. Ultimately, my analysis may help to explain why Tolstoy included such an extended, seemingly peripheral and non-consequential depiction of the hunt in his vast novel about Russia's multivalent encounter with the West.⁴²

Figure 3 Rudolf Ferdinandovich Frenz, Grand Duke Vladimir Alexandrovich of Russia on a wolf hunt (1890s). Image copyright Lebrecht Music & Arts

My method will be to first sketch the principle sources that will provide a context for a close reading of Tolstoy's borzoi hunting scenes, then present Tolstoy's portrayal of the wolf hunt in discrete stages, stepping back into this broader context at each stage. This approach will involve some jerkiness as I maneuver between Tolstoy's text and this larger context but will ultimately yield a nuanced understanding of some of the most important issues at play. At the chapter's end I will suggest some recurrent patterns that underlay Russian depictions of borzois, wolves and those Russians who hunted with them, contrasting them to the very different paradigms of English fox hunting. I will also contrast Nikolai's ruthlessness as a wolf hunter with the compunction that overcomes him in a later battle scene when he confronts not a wolf but a human adversary. Taken as a whole, the chapter will set the stage for exploring the changing attitudes toward wolves that informed gun hunting and wolf eradication efforts later in the century in Chapter 2.

Among the sources mentioned above I will devote particular attention to a loosely memoiristic account of 1859 by Tolstoy's little-known contemporary Yegor Eduardovich Driansky (~1812-1872) entitled *Notes of a Hunter with a Small Leash of Hounds*.[43] Driansky, who was about a decade-and-a-half older than Tolstoy (1828-1910), was born in Ukraine but spent his adult years in Russia and published in some of Russia's leading journals in the 1850s and 1860s as a protégé of the playwright A. N. Ostrovsky. Despite his substantial literary talents, his career foundered in part because of his difficult personality, which led to misunderstandings with editors and others and—although he was a member of the gentry—he spent the last years of his life in impoverished circumstances. Portions of Driansky's most significant work, his *Notes of a Hunter*, first appeared in installments in the journals *The Muscovite* and *Library for Reading*. In 1859—just two years before the emancipation of the serfs and half a decade before Tolstoy's first drafts of *War and Peace*—the work was published separately as a supplement to the newly established monthly journal *The Russian Word*.

Driansky's *Notes*, which have been neglected despite literary qualities that rival those of better-known contemporaries who wrote autobiographical accounts of hunting combined with expert observations and ruminations about wildlife and nature such as S. T. Aksakov, will help to elucidate the hunt as a means of characterization and the interaction of the social, the domestic, and the wild at play in Tolstoy's portrayal. His first-person narrative, which is nearly 200 pages long, conveys the experiences of a provincial nobleman who starts out as a proponent of gun-hunting but is introduced by a well-meaning friend—like Rostov a count—to the institution of hunting with hounds. He recounts his growing understanding of the nuances of the blood sport as he accompanies this mentor, Count Aleev, and his entourage on a number of hunting expeditions for wolves, foxes, and hares in European Russia. Over time he becomes familiar with an extensive range of hunting jargon (for example, the terms used to describe wolves of different ages), which he explains to the reader, as well as to the finer points of how to train both scent hounds and borzois, successfully stage a wolf hunt, and judge the qualities of both the dogs and the huntsmen who train them and accompany the gentry on their hunts. Even for his contemporaries and now modern readers with expertise in the period, Driansky's use of specialized hunting vocabulary presents challenges, although it lends his work an air of authenticity and provides the reader with a sense of discovery.[44] Tolstoy deploys such specialized vocabulary more sparingly but his text conveys the same sort of intimate knowledge of borzoi hunting, as we shall see. Among his contemporaries and subsequent generations, Driansky's portrayal was viewed as extremely accurate and precise, and referred to as a reliable source of insight into the institution of borzoi hunting.

Three guides to hunting with scent hounds and borzois will inform this comparative analysis and lend insight into the larger issues at play. The first of these was written by N. Reutt, who was the editor of the same *Journal of Horse Husbandry and Hunting* that published Volunin's 1842 account of Brigadier General Boriatinsky's encounter with a wolf. Reutt's *Hunting with Hounds* appeared in 1846.[45] Reutt lived and hunted in the Kostroma Province, the capital of which was located about 200 miles northeast of Moscow. He stated at the outset that his mission was to clarify and elucidate the best methods of hunting with borzois and scent hounds, given the dearth of previous such guides in Russia. His work spans two volumes each more than 200 pages long. The first is devoted to the breeding, evaluation, and care of borzoi hounds. The second addresses horses, scent hounds, and techniques for hunting foxes, hares, and wolves. As the borzoi guide most temporally contiguous with the two literary works in question, Reutt's compendium includes numerous references to classical treatises on hunting ranging from Aristotle to Plutarch, hunting tomes of the middle ages, and Western European treatments of the hunting theme. Although he professes himself an Anglophile, Reutt boasts of Russia's vast areas of forest and wilderness in comparison with England's tamed landscapes and manicured parks.[46]

Precisely forty years after Reutt's *Hunting with Hounds*, P. M. Machevarianov's *Notes of a Hunter with Hounds from the Simbirsk Province* appeared as a supplement to the *Journal of Hunting* in 1876.[47] Machevarianov (~1804-1880) was a wealthy landowner who retired from military service to a country estate about 300 miles southeast of Moscow. An ardent hunter of substantial means, he maintained as many as 300 hounds in his kennel and was renowned for his hunting expertise and the quality of his prize-winning borzois (although some contemporaries noted that his borzois were not typically the most ferocious and were better suited to fox and hare than wolf hunting). In the foreword to his *Notes* Machevarianov bemoaned his contemporaries' ignorance of proper borzoi breeding and hunting techniques but praised Driansky's work as an exception.[48] His volume is considered to be the single most authoritative nineteenth-century source on Russian borzois. In addition, it is infused with his attitudes as a member of the nobility toward issues of class and national identity in the aftermath of the emancipation, as well as the conviction that Russia's borzois and wolf hunting with them represent a quintessential aspect of national identity, as exemplified in the epigraph that opened this chapter. In the passage Machevarianov associates genuine "Russianness" with hunting prowess and specifically the ability to dominate and kill a mature wolf with a dagger or capture it alive. He dissociates these qualities from the foreign influences embodied in Russia's capital cities of Moscow and especially St. Petersburg. He links these aspects of an archetypal masculine national character directly with Russia's landscape and wildlife.

The third major work on which I will draw for context is P. M. Gubin's *Complete Guide to Hunting with Borzoi Hounds*, published in 1891.[49] Among the three guides to borzois, Gubin pays the most sustained attention to wolf hunting: he draws on thirty years of experience hunting wolves with borzois and devotes dozens of pages to the topic.[50] He precisely enumerates the roles and required qualities of the various participants in the hunt, ranging from the chief huntsman to the master of hounds, demarcates the calls and commands to be employed—complete with musical notation—and provides extremely granular detail about all aspects of hunting with borzois. At the same time—as in the case of Machevarianov but in a different vein—Gubin's portrayal of borzoi wolf hunting reveals deep class prejudices alongside his own ethical concerns about issues ranging from trespassing to the use of strychnine, which he vehemently opposes. For example, he repeatedly blames the peasantry for abetting livestock losses to wolves by allowing their domestic animals to trespass, which he claims leaves them unprotected from wolves while also demonstrating the peasants' lack of respect for private property.[51]

Driansky's work is the most explicitly literary among this group of texts, yet each transcends the generic boundaries of guidebook or manual through the inclusion of personal anecdotes and authorial interjections like the brief examples above. These reveal the authors' class affiliations and interrelated attitudes toward nature and culture in ways that lend insight both into Tolstoy's novel and also changing attitudes toward the role and meaning of borzoi wolf hunting in Russia as it evolved over the century. In this sense, they are reminiscent of Aksakov's guides to fishing and bird-hunting, which describe techniques and skills useful for the fisherman or bird hunter but also devote substantial attention to the author's ruminations on nature, particular species, and ethical issues that reflect his class sensibilities as well as his zoological expertise.[52]

* * *

Tolstoy's extended description of a wolf hunt in War and Peace is one of the lengthiest passages involving human interactions with nature in the novel. Yet the four chapters devoted to the wolf hunt, along with the subsequent pursuit of a fox and hare, devote as much attention to the hunt as a social institution and to the rivalries that underlie the hunters' interactions as to the wild prey (wolves, fox, and hare) they pursue and to the domesticated animals (borzois, scent hounds, and horses) who mediate their relationships with their quarry. Understanding these interwoven aspects of Tolstoy's portrayal requires knowledge of each of these aspects of the hunt. Together, they lend insight into Tolstoy's characterization of the old Count Rostov, his son Nikolai (an ardent but relatively inexperienced hunter), their fiercely independent huntsman Danilo, the young countess Natasha as a feminine intruder into this masculine domain, and the neighbors with whom they hunt—especially the eccentric character whom Tolstoy calls "Uncle." In addition, Tolstoy's striking portrayal of the old wolf reveals an interesting ambivalence that may hint at his later revulsion of feeling about hunting, which I will explore more thoroughly in Chapter 4.

The hunt, which Tolstoy depicts primarily from Nikolai's perspective, also serves as a pivotal moment in Tolstoy's characterization of Nikolai as a young nobleman who is attempting to regain and assert a wounded sense of masculinity. Along with his experiences in battle and a calamitous gambling loss, Nikolai's role in capturing a mature wolf at the culminating moment of the hunt represents a momentous event in his own process of maturation. He has returned to his ancestral home between military campaigns during which he also suffered a catastrophic loss of 43,000 rubles at cards to a fellow military officer, the ruthless cardsharper and duelist Dolokhov. During this earlier confrontation over cards Dolokhov sadistically manipulates and exploits Nikolai through the latter's naïveté and allegiance to the aristocratic codes that underlie such notions as the "debt of honor."[53] Dolokhov almost certainly cheats as he deals the cards, taunting the passive and submissive Nikolai as "a cat plays with a mouse."[54] Later in the novel Dolokhov will also participate in an attempt to seduce and abduct Nikolai's sister, Natasha. He is a predator in various senses, and one may view Nikolai's eventual triumph over the mature wolf as a symbolic retrieval of the honor that Dolokhov has imperiled.[55] Moreover, Tolstoy directly compares Nikolai's perceptions and actions during a battle scene later in the novel with his conduct during the wolf hunt, as we shall see. These interrelated scenes demonstrate how closely the semiotic sphere of borzoi hunting was interrelated with other areas of male gentry behavior such as gambling and dueling. Each of these behavioral spheres allowed young gentlemen like Nikolai to affirm (or question) the codes of honor and displays of masculine aggression that underpinned male gentry identity, which came under increasing scrutiny as the century progressed.

At the same time, the hunt operates on an entirely different level that affords another sort of redemption. It provides the characters and Tolstoy's readers with an opportunity to penetrate briefly or at least to experience close proximity to a mysterious and mesmerizing world of nature that lies beyond human culture and social convention. Natasha emits a piercing scream of joy at the end of the hare coursing, which corresponds to Nikolai's rapture as his borzois overwhelm the wolf. Wildly out of place in any other context, evoking yet unlike a wolf's howl, her scream captures this sense of freedom from convention and the delight that all the hunters take in their borzois' pursuit and bloody capture of the swift prey, which they follow on their galloping horses. Although they are intruders into this natural world, the hunters participate in it intensely through the mediation of their horses and hunting dogs, whose exquisite senses amplify and intensify the hunters' own perceptions, and who fight the wolf and other prey with only their teeth as weapons. In crucial sections of the narrative, Nikolai seems to be experiencing the world vicariously through the senses of his favorite scent hounds and borzois, somewhat like Tolstoy's protagonist Levin would experience the marshland and gamebirds through the perceptions of his pointer Laska in the novel *Anna Karenina* a decade later.[56]

Tolstoy himself was an avid hunter who pursued wolves from the 1850s through the 1870s, prior to renouncing hunting and espousing vegetarianism at about the age of fifty in connection with his acute spiritual crisis of the late 1870s and early 1880s.[57] Several of his letters from 1859, when he was in his early thirties, refer to recent wolf hunts including some at the family's estate of Nikolskoe-Viazemskoe with his acquaintance from military service in the Caucasus, I. P. Borisov. It appears that Tolstoy hunted wolves primarily or exclusively with dogs rather than as a gun hunter. In an August 29, 1859 letter to Borisov Tolstoy mentioned that he had just returned from a hunting trip to the Kashirsky and Venevsky districts where he had taken two of five wolves and three of six foxes killed by the hunting party as a whole.[58] In another letter of October 1, 1859 from Nikolskoe-Viazemskoe to his brother Nikolai, from whom he would inherit the estate in 1860 on the latter's death, Tolstoy mentioned that he had taken two wolves and eleven foxes, while Borisov had taken one wolf, in recent hunts. "All this time I've done nothing but hunt and am content with that," he added.[59] In an October 24, 1859 letter addressed to both the poet A. A. Fet and Borisov, Tolstoy thanked Borisov for lending him scent hounds, his huntsman, and a gelding. He added that he would keep them until the snowfall.[60] Tolstoy hunted wolves as late as 1877, when he spent part of August hunting wolves on the estate of D. D. Obolensky.[61]

Tolstoy's intimate familiarity with wolf hunting informed his depiction of the hunt in *War and Peace*, which is extremely accurate in all respects down to the precise use of hunting jargon, the staging of the hunt and composition of the hunting party, the roles played by various participants, and the way in which his dramatis personae's knowledge and attitudes toward hunting, their dogs, and their prey serve as signifiers within this semiotic field. Tolstoy's representation both cleaves to and diverges from the mythical archetypes presented in Volunin's account. His creative distortions and variations lend his portrayal an idiosyncratic immediacy, while also reflecting his own nuanced understanding of the underlying cultural institution as well as his deep attachment to nature and to hunting as an especially intense and meaningful way of interacting with the natural world.

The first chapter of the hunt opens with an extended description of the changes that autumn has brought to the flora and fauna of the Rostovs' estate. Nikolai's perception of these seasonal changes connects directly with his knowledge of the conditions that make for good hunting, which he—as an ardent but still relatively inexperienced huntsman—has begun to understand. These seasonal changes correspond to metamorphoses in the three sorts of prey they will seek:

> It was already turning winter, morning frosts gripped the earth moistened by autumn rains, the winter wheat was already tufting up and stood out bright green against the strips of brownish, cattle-trampled winter stubble and pale yellow summer stubble with red strips of buckwheat. The hilltops and woods, which at the end of August were still green islands among the

black winter croplands and stubble, had become golden and bright red islands amidst the bright green winter crops. The hares had already half shed their summer coats, the fox cubs were beginning to disperse, and the young wolves were bigger than dogs. It was the best time for hunting.[62]

Every detail of this description, filtered through the eager hunter Nikolai's perceptions, corresponds to the practices that governed borzoi hunts, which typically took place in the fall when the crops had been gathered and the wolf pups were large enough to provide something of a challenge but still lingered close enough to their dens to keep the mature wolves in a pack anchored in the vicinity. Tolstoy contrasts the bare "cattle-trampled" fields, shorn of their crops during the harvest, with the wooded "islands" (*ostrovy*) or copses that provide cover for their intended prey. It was important, of course, that hunters on horseback avoid trampling fields prior to the harvest, and hunts also most often took place when the thick vegetation of summer had thinned but before deep snow made pursuit too difficult.

While the Rostovs' wolf hunt takes place primarily on their own extensive estate, the following pursuit of a fox and hare take them farther afield in the company of their neighbors the Ilagins. It's worth mentioning in this connection that Machevarianov claims that Russian property owners would usually gladly give permission to gentlemen hunters to hunt on their land. He contrasts this with the more legalistic customs of the West in an effusive testament to Russian patriotism, which he links with the granting of hunting access to fellow landowners:

> Material calculations, which are so prevalent among Western peoples, are entirely foreign to our Russian landowners . . . In our provinces a hunter need only drop a hint to receive not only permission but a heartfelt invitation from a landowner or manager; this is why the Russian hunter always boasts to foreigners about the right of free untrammelled hunting in all corners of Russia."[63]

Machevarianov's effusiveness seems overstated (and ironic in the context of Tolstoy's novel, as an argument later ensues over whose land the fox is taken on), and one gathers that his emphasis on a patriotic tradition of shared access for hunting applies exclusively to the gentry, but it is nevertheless noteworthy. It also reflects the much larger tracts of undeveloped land and forest that remained in Russia in comparison with most of Western Europe.

On the morning of the hunt both Nikolai and his restless borzois, who have been kept idle for the previous three days in preparation, sense that conditions are perfect:

> The only movement in the air was the slow movement from above to below of descending microscopic drops of mist or fog. Transparent drops hung on the bare branches of the garden and dripped onto the just-fallen leaves. The soil in the kitchen garden, glistening wet and black as poppy seed, merged in the near distance with the dull and damp curtain of the mist. Nikolai went out to the wet, mud-tracked porch; there was a smell of fading leaves and dogs.[64]

As he and two of his borzois linger in the garden amid the moist and scent-laden air, his huntsman Danilo comes around the corner of the house:

> "Hal-loo!" Just then came that inimitable hunting call, which unites in itself the deepest bass and the highest tenor; and around the corner came the head kennelman and huntsman, a wrinkled old hunter, his gray hair cut round in Ukrainian fashion, a hooked hunting whip in his hand, and with that expression of independence and scorn for everything in the world that only hunters have. He took off his Circassian hat before his master and looked at him scornfully. This scorn was not offensive for the master: Nikolai knew that this Danilo, who scorned everything and was above everything, was still his serf and his hunter.
> "Danilo!" said Nikolai, timidly sensing that, at the sight of this hunting weather, these dogs, and his hunter, he was already being seized by that irresistible hunting feeling in which a man forgets all his former intentions, like a lovesick man in the presence of his beloved.[65]

Tolstoy's narrative, linked with Nikolai's perceptual point of view, immediately privileges the huntsman Danilo. Embodied in his rich multi-toned voice, his fiercely independent character transcends the confines of domesticity symbolized by the house and garden, of which Nikolai is the master. Nikolai's timidity captures this reversal of authority between master and serf, as the roles of the coming hunt begin to establish dominance. Danilo's undomesticated demeanor, with his Ukrainian haircut, Circassian cap, and hunting whip ready at hand recalls the fiercely independent Cossacks that Tolstoy had described in his 1863 novel *The Cossacks*, and especially the Cossack hunter Eroshka. Like Eroshka, Danilo is a liminal figure who mediates and facilitates Rostov's and the other hunters' encounters with nature, just as Eroshka guides Olenin during his forays into the Caucasian wilderness.[66]

Tolstoy's portrayal of Danilo may seem overstated or idiosyncratic in its emphasis on his independence and arrogance (he is the count's serf, after all). Viewed in the context of portrayals of huntsmen in other nineteenth-century sources, however, it is less surprising. The role and responsibilities of the chief huntsman of a large hunting establishment in this era were diverse and required considerable expertise and experience. An especially accomplished and trusted individual could be responsible for several dozen or even a hundred or more scent hounds and borzois, as well as the assistant huntsmen, kennel attendants, whippers-in, and others who tended to the dogs and kept track of local game populations, as well as hunting opportunities farther afield.[67] He also played a central role in training the scent hounds and borzois, although this role could be divided between two or more individuals as well. This required a combination of qualities that were highly valued and respected. Sometimes such men were hired for pay—as became the norm after the emancipation—but prior to the emancipation they were typically selected from among a landowner's serfs. In either case, however, the primacy of the head huntsman was understood and respected.

Machevarianov, writing in 1876, emphasizes the importance of selecting capable individuals as huntsmen and especially warns his readers against employing household servants in these roles. He argues that one is much better off paying fairly for a capable and dedicated huntsman than losing the money at cards or spending it on expensive gifts of jewelry for women, as hunting is better for one's character than engaging in these other masculine pastimes.[68] He encapsulates the responsibilities of the head huntsman (*lovchy*) as follows:

> The head huntsman is the primary overseer of the entire hunt, who answers for any mishaps. He is distinguished by his exemplary behavior, his sobriety, his keen-wittedness, and his complete mastery of the coursing of the kennelman and borzoi handler. All the hunters answer entirely to him.[69]

Gubin also affirms the chief huntsman's stature and responsibilities. In addition, he emphasizes the importance of a sonorous and powerful voice with good tonal range, capable of carrying over long distances and consonant with the "music" of the hounds (this helps to explain Tolstoy's emphasis on Danilo's penetrating voice). He even provides detailed musical notation for the various hunting calls used in wolf, fox, and hare hunting.[70] In his more literary account Driansky describes encountering several chief huntsmen, each of whom has his own particular personality and skills. He especially emphasizes the importance of passion for the hunt and daring in fulfilling this role in ways that are strikingly reminiscent of descriptions of noble pursuits traditionally associated with the gentry (rather than their servants), such as gambling and other forms of risk-taking. After dismissing those who fall short of expectations, he extolls the qualities of an ideal huntsman:

> But a chief huntsman by nature, by calling, by desire—that's another thing and something new and unimaginable for those who haven't themselves experienced an example drawn from real life of what a man is capable of and what he can do out of ardor, passion, desire. A huntsman by calling is a rebel, a daredevil—stake life on a copeck! He's a man who astounds others, a man who of his own will, out of longing has consigned himself to labor, to risk, to trial, to torture . . . By his very nature he must be unlike other men; he's a man apart, hardy, made of iron![71]

These portrayals by Tolstoy's contemporaries are striking, given that they ascribe qualities of daring [*udal'stvo*] and independence that were more typically associated with members of the male nobility instead to the serfs or paid employees who served them. These accounts and Tolstoy's characterization of Danilo (and the earlier Cossack hunter Eroshka) imply that a shared love for hunting provided an opportunity for mutual respect between the gentry and their huntsmen. In addition, the dedication and expertise that a skilled and passionate huntsman could provide to his master, prior to the emancipation, or employer in the post-emancipation era, offered a vicarious intimacy with the natural world that the latter might not be able to achieve on his own.

Along these lines, a primary responsibility of huntsmen was to keep appraised of local game populations, which often involved interacting with the local peasantry on behalf of the gentry. This included locating wolf dens during the summer, then keeping an eye on them until the fall to ensure that the wolves remained in the same or a nearby location. Gubin describes in detail how wolf dens should be located between June 15 and September 1. He provides a list of seventeen questions that huntsmen could pose to locals in order to pinpoint the location of a den. These demonstrate his considerable skepticism and distrust of peasants' own motivations and abilities to accurately assess wolf behavior, which he ascribes partly to their desire for vengeance on wolves that have killed domestic animals they have let trespass unattended onto neighboring lands. He asserts that this will lead them to provide false or misleading information—for example claiming that wolves have a den much closer to their villages and pasturelands than it is in reality:

> In the majority of cases local inhabitants always lie about wolves, but an experienced hunter mustn't allow this to cloud his judgement, but on the contrary must always carefully ask around about wolves, for if the wolves' den is not close to a village, then one can always assume that it's somewhere four or five versts distant if it's a case of frequent, nearly daily, loss of livestock in a certain place.[72]

Gubin's advice is interesting on a number of levels. For one thing, it corroborates other accounts of wolf dens being kept under surveillance over the summer months and prior to the fall hunting season, rather than targeted while the pups were younger and more vulnerable. This indicates that many gentry hunters "reserved" wolves to provide satisfying hunting later on, rather than simply killing them as expeditiously as possible on the discovery of their dens, and somewhat undercuts their claims to be protecting the interests of the peasantry. In addition, Gubin asserts elsewhere that the peasants bear partial responsibility for livestock losses through their own tendency to let their animals wander.[73] Given the tensions that arose over grazing rights in the wake of emancipation and the division of land between the former landlords and peasantry, this is a striking example of gentry self-justification.[74]

For now, let us return to *War and Peace* as preparations for the hunt continue. Danilo tells Nikolai that he has sent his assistant Uvarka to pinpoint the current location of the wolves: "I sent Uvarka to listen at dawn," his bass said after a moment's silence. "He says she *transferred* them to the Otradnoe reserve. They were howling there."[75] Soon the master huntsman and his assistant present themselves to make final plans for the hunt:

> Five minutes later Danilo and Uvarka were standing in Nikolai's big study. Though Danilo was of small stature, seeing him in the room produced an impression similar to seeing a horse or bear standing there amidst the furniture and accessories of human life. Danilo felt it himself and, as usual, stayed near the door, trying to speak softly, not moving, so as not to break anything somehow in his master's room, and trying to say everything as quickly as possible and get out into the open, from under the ceiling to under the sky.[76]

To Danilo's further discomfort and Nikolai's annoyance, his sister—the seventeen-year-old countess Natasha—and their younger brother Petya run into the study and insist on joining the hunt. Natasha rebels against traditional gender roles: she calls the hunt "my greatest pleasure" and orders a huntsman to bring her own pack of dogs as well.[77] This corresponds to Tolstoy's further development of her character through the hunting sequence, which serves to emphasize and privilege her appropriation and expression of a uniquely Russian femininity, as well as her disruption of masculine codes.[78] Natasha's possession of her own pack of hunting dogs reflects the scale of the Rostovs' hunting establishment, as only women of the high aristocracy and royalty typically had hunting borzois.[79]

The full size of the hunting party becomes evident as it sets out with Nikolai at its head on his chestnut Don stallion. It resembles a military formation setting off for battle, which is consistent with Nikolai's role as an army officer home between campaigns:

> In all, fifty-four hounds were led out under six kennelmen and whippers-in. Besides the gentlemen, there were eight borzoi handlers, around whom roamed more than forty borzois, so that, with the gentlemen's packs, there were about a hundred and thirty dogs in the field and twenty mounted hunters.
>
> Each dog knew its master and its name. Each hunter knew his task, place, and purpose. As soon as they went beyond the fence, everybody, with no noise or talk, spread out evenly and calmly along the road and field leading to the Otradnoe woods.[80]

Gubin, writing in 1890, provides very specific numbers by which we can evaluate the size of the Rostov hunting party in historical terms. He notes that a hunt should consist of an absolute minimum of eighteen scent hounds with a head kennelman and two assistants, as well as at least five leashes of borzois with four to a leash. The largest hunt, Gubin continues, should not have more than forty scent hounds, with a kennelman and three assistants, and twelve leashes of borzois with three to a leash. Gubin mentions, however, that in earlier times hunts sometimes consisted of as many as five hundred hounds, although these would typically be broken into separate packs.[81] Reutt, writing almost half a century prior to Gubin and much closer to the era that Tolstoy was portraying in the novel, presents a less quantitative prescription and one fascinating in its own terms as he focuses on the number of scent hounds and their "voices." After first advising his readers to maintain a hunting establishment that is within their means so as not to end up "without forests or fields," he continues:

> The number of leashes must be in harmony with the locale because it's just as unpleasant to hear a large pack of hounds in a small wood as it is to hear a large orchestra in a small room; and then a small pack in a large forest, in dense groves, calls to mind the existence of a man of small means out in high society.[82]

Reutt's prescription is especially interesting as it emphasizes the importance of the sound of the pack, which was a crucial aspect of hunting with scent hounds in Russia, just as it was for foxhunting in England. His use of similes based on the mores of high society and musical performances reflects his presumed readership and his attempt to convey his point in mutually intelligible terms. Moreover, his emphasis on not spending beyond one's means by maintaining an overly large hunting establishment corresponds to the situation portrayed by Tolstoy, in which the Rostovs' maintenance of a large kennel is one of the factors that is contributing inexorably to their financial run.

As they approach the woods in which the wolves' den has been located the Rostov party encounters five additional horsemen with their own dogs, among whom is "Uncle," their distant and much poorer relation and an inveterate hunter. He warns them that their neighbors, the Ilagins, are intent on snatching the wolf cubs out from under them (this was a common practice and concern, which occurs repeatedly as a motif in Driansky's *Notes* as well). Nikolai invites the newcomers to join them. As they approach the woods where the den is located the various characters reveal greater or lesser mastery of the hunt as an arena for the display of masculine competence. Nikolai, for example, is quick to put his sister Natasha in her place when she refers to one of his favorite harriers (a small bloodhound) simply as a "dog" (*sobaka*): "'First of all, Trunila's not a dog, he's a bloodhound (*vyzhlets*),' thought Nikolai, and glanced sternly at his sister, trying to make her feel the distance that should separate them at that moment. Natasha understood it."[83]

As Danilo prepares to unleash the hounds within the forest, Nikolai and "Uncle" help to arrange where the different hunters and borzois should be stationed—a task that requires skill and experience in anticipating the most likely escape routes of the wolves. After placing Natasha in a location "where nothing could ever run out," Nikolai stations himself with his old borzoi Karai in a choice spot: "Karai was an ugly and whiskery old he-dog, famous for having gone alone against a seasoned wolf."[84] Like Boriatinsky's Beast, the name Karai connotes aggression, as it derives from the verb *karat'* (to punish or chastise). In general, borzois' names often derived from verbs of this sort or connoted strength, aggression, or savagery by association. Gubin provides an extensive list of habitual names for borzois in his *Guide*. Examples include *Zlodei* (villain or scoundrel), *Khishchnyi* (predatory or rapacious), *Pozor* (shame or disgrace), *Satana* (Satan), *Demon* (Demon), and a host of others of the same ilk. Significantly, many of these terms also regularly appeared in descriptions of wolves.[85]

For Russian borzoi hunters a mature wolf was the ultimate quarry. Such a large animal's speed, strength, endurance, and ferocity required careful positioning of hunters on their horses with leashes of three or four borzois ready at the outlets (*lazy*), or funnels, from which the animals would most likely attempt to escape the large packs of scent hounds that were sent into the copses and larger woods to drive them out. Wolf hunts required the boldest and strongest borzois, which had to be let loose at precisely the right moment as the fleeing wolves came into sight (borzois rely primarily on sight not scent), almost as a falconer might release a bird of prey at a quickly-moving target. If the borzois were unleashed at the wrong moment they could either fail to catch a running wolf (they excelled at sprinting rather than endurance running) or run past it on the wrong trajectory, missing their quarry. In addition, younger or more timid borzois could be daunted and shrink away as they closed in on a wolf; this made it extremely important to have older and more experienced borzois in the mix.

Gubin elucidates these principles in his *Guide*, which devotes dozens of pages to borzoi wolf hunting. He notes that hunters should wait at their chosen or assigned spots in absolute stillness while leaning close to the neck of their horses and speaking no louder than a whisper to calm their leashed borzois while awaiting the sudden appearance of the prey.[86] He also emphasizes the chief huntsman's role in placing hunters according to their abilities.

> The coursing of the borzoi handlers is determined by the chief huntsman according to the audacity, keen-wittedness, experience and quickness of those handling the borzois. This is because the ability of the borzoi leashman to point out the beast, to unleash the dogs upon it, and to retrieve the beast from them expeditiously play an especially significant role in wolf hunting.[87]

As Tolstoy's portrayal continues, the characters are judged by their competence and skill in embodying these abilities. Tolstoy first devotes two pages to a humorous portrayal of Nikolai's father, the old Count Rostov. The count waits with his attendants and his "three wolfhounds, spirited but grown fat, like their master and his horse" in his appointed spot—a choice one in light of his status—like a "child made ready to go for walk."[88] We learn that "though not a hunter at heart, [he] had a firm knowledge of the rules of hunting."[89] In his bemused state, however—likely aided by the silver tumbler of spiced brandy and half bottle of Bordeaux he has downed prior to the hunt—the old count breaks a cardinal rule of borzoi wolf hunting. He and his attendants—who are busying themselves by complementing his son's horsemanship and hunting prowess as he partakes of some snuff—fail to anticipate the emergence of a wolf from the forest and to unleash the borzois at the right moment.

This precipitates one of the most idiosyncratic moments in Tolstoy's characterization of the wolf hunt, as he implicitly compares the wolf to the aging count:

> The count and Semyon leaped out of the bushes and to their left saw a wolf, which, swaying softly, was moving at a gentle lope to the left of them, towards the same bushes by which they were standing. The angry dogs squealed and, loosed from their leashes, raced toward the wolf past the horses' legs.
>
> The wolf slowed his flight, turned his big-browed head towards the dogs awkwardly, as if suffering from angina, and, swaying just as softly, leaped once, twice, and, with a wag of his tail, disappeared into the bushes.[90]

In the next instant the pack of scent hounds appears in hot pursuit, followed immediately by Danilo on his sweat-bathed horse:

> When he saw the count, his eyes flashed lightning.
>
> "A-----!" he cried, raising his whip threateningly at the count. You b-----ed the wolf! . . . Some hunters!
>
> And as if not deeming the abashed, frightened count worthy of further conversation, he whipped the hollow, wet flanks of his brown gelding with all the anger he had prepared for the count, and raced after the hounds . . . But the wolf got through the bushes, and not one hunter intercepted him.[91]

The extended passage devoted to the old count, which culminates in his serious lapse of good sportsmanship, seems intended by Tolstoy in part for humorous purposes. In addition, it highlights the inversion of social hierarchy that allows the serf huntsman Danilo to castigate the count so vehemently, brandishing his whip at him, for his unforgivable lapse.[92] The passage also allows for Tolstoy's unusual characterization of the fleeing wolf, which so unexpectedly casts it in a sympathetic light by emphasizing its awkwardness and discomfort. It is hard not to recognize an implicit parallel with the age and approaching mortality of the count, particularly given that he and his attendants are the ones who witness the wolf in this light.

After presenting the initial stages of the hunt from the vantage point of the old Count Rostov and his retainers, Tolstoy shifts to Nikolai's very different viewpoint, which he presents in much more intimate terms through a combination of indirect discourse and direct quotation of Nikolai's thoughts. This emphasizes the dramatic difference in age and outlook between father and son, as well as Nikolai's greater emotional investment in the hunt's outcome. For the old count, who presumably has been on many such hunts, the day represents an opportunity for a pleasant outing and reminiscences with his devoted servants. For Nikolai, on the other hand, the hunt represents an unprecedented opportunity to test himself by confronting the Russian borzoi hunter's most elusive and sought-after quarry, to redeem himself after his recent gambling loss, and to quell his abiding state of anxiety over his family's waning fortunes.

The shift to Nikolai's viewpoint, which opens a new chapter, begins with close attention to the sounds of the hunt. Listening intently to the hounds' and kennelmen's voices and cries, Nikolai is able to detect that the hounds have broken into two packs, that there are both young and old wolves, and that something has gone awry. His ability to interpret the "music" of the hounds to draw these conclusions indicates his growing expertise as a hunter. Reutt emphasizes the beauty and musicality of the chorus of hounds in his 1846 guide but also the insight an experienced hunter can gain simply from hearing the hounds in the distance:

> The vocal congruence of the pack or the harmony of the voices during the chase possesses the same beauty in hunting as does refined coloration in art. A well-matched pack must cover six octaves, with the highs and lows of each octave. The quickness and slowness of the pack's throat utterances, just like their pace, expresses the greater or lesser freshness of the scent trail and adjusts in conformity with the type of animal they are

pursuing... A hunter who is familiar with this harmony and the sound of the voices of his dogs can easily tell what sort of prey they are chasing.[93]

For Nikolai, the hunt represents a moment pregnant with the possibility of encountering and subduing a mature wolf. His anguished hope that the wolf will come his way leads him to pray fervently:

> Several times he addressed God with a plea that the wolf come out at him. He prayed with that passionate and guilty feeling with which people pray at moments of strong agitation arising from insignificant causes. "What would it cost You?" he said to God. "Do it for me! I know You are great, and it's a sin to ask it of You, but, for God's sake, make it so that the old wolf comes my way and Karai, before my uncle's eyes, gets a death grip on his throat... No, such luck is not to be," thought Rostov... "I'm always unlucky, in cards, in war, in everything." Austerlitz and Dolokhov vividly but fleetingly flashed in his imagination. "If only once in my life I could chase down a seasoned wolf, I'd ask for nothing more!" he thought, straining his hearing and sight, looking to the left and then to the right, and listening to the smallest nuances in the sounds of the chase.[94]

Here the psychological and even spiritual interplay between Nikolai's traumatic experiences in battle, his loss at cards to Dolokhov, and his hopes to take a mature wolf become explicit. The stakes of the hunt replace those of the field of battle and the card table as Nikolai seeks redemption on this different but closely related characterological playing field. In this sense the semiotic spheres of battle, gambling and hunting—all of which allowed young nobleman to display their bravery against a backdrop of chance and luck—presented parallel opportunities for them to earn distinction or disgrace, as Tolstoy knew well.

Nikolai can scarcely believe his eyes when—after half an hour of his rapt attention to the ebb and flow of the hunting horns and the baying of the hounds—a large adult wolf appears from the woods running toward him:

> "No, it can't be!" thought Rostov, sighing deeply, as a man sighs at the accomplishment of something he has long-awaited. What was accomplished was his greatest happiness—and so simply, without noise, without splendor, without portent. Rostov could not believe his eyes, and this doubt continued for more than a second. The wolf ran on and jumped heavily over a hole that lay in his path. He was an old beast, with a gray back and a well-stuffed, reddish belly. He ran unhurriedly, obviously convinced that no one could see him...

"Loose them, or not?" Nikolai was saying to himself all the while the wolf moved towards him, drawing away from the woods. Suddenly the wolf's entire physiognomy changed; he shuddered at the sight of human eyes, which he had probably never seen before, directed at him, and turning his head slightly towards the hunter, stopped—go back or go on? "Eh, it makes no difference, I'll go on!" He seemed to say to himself and started forward, not looking around now, at a soft, long, free, but resolute lope.[95]

As readers we cannot be sure whether this is the same wolf encountered by Nikolai's father, as typically both the male and female of the dominant adult pair would stay near the den site along with the previous year's now adolescent pups and the current year's litter. Moreover, Tolstoy observes that the wolf has likely never before encountered human eyes directed on him. In either case, Tolstoy's portrayal of the mature wolf is consistent with his earlier representation and equally idiosyncratic. He emphasizes the wolf's "well-stuffed, reddish belly" and initial complacency that he is unobserved. For his part, Nikolai has been straining his senses of hearing, sight, and smell, which are now fully engaged on the wolf. The wolf recoils with a visceral shock at sensing Nikolai's human gaze on him, which he seems to experience almost as a physical sensation. Tolstoy's decision to verbalize the wolf's thinking is also striking. Presenting animals' thoughts either indirectly or in quotes was a common but sometimes critiqued tactic in literature of the day, as I will explore further in close readings of wolf-centered narratives in Chapter 4.

Nikolai acquits himself much better than his father. He unleashes his borzois and begins a furious pursuit of the wolf on horseback, urging them on with the cry "U-liu-liu-liu!, U-liu-liu!," which translates roughly as "Take it down!"[96] Tolstoy devotes half a page to the pursuit, as Nikolai witnesses his three borzois—Milka, Lyubim, and Karai—attempt to catch the swift wolf while avoiding his rending jaws.

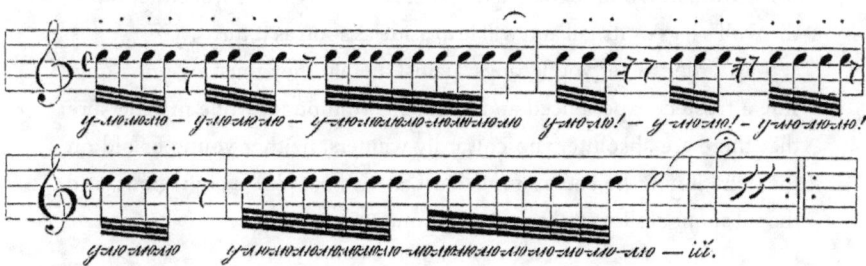

Figure 4 Musical notation for a borzoi hunter's call to attack a wolf. From P. M. Gubin, *Polnoe rukovodstvo ko psovoi okhote v trekh chastiakh* (Moscow: Tipo-litografiia Morits Ivanovich Neibiurger, 1890).

The red Lyubim overtook Milka, precipitously threw himself at the wolf, and seized him by the hindquarters, but in the same second became frightened and jumped over to the other side. The wolf crouched, clapped his teeth, got up again, and loped on, accompanied at two yards' distance by all the dogs, who would not go nearer to him . . . A young, lanky, brindled dog, unknown to Nikolai, from another pack, flew swiftly at the wolf from the front and nearly bowled him over. The wolf got up more quickly than might have been expected of him, rushed at the brindled dog, snapped his teeth—and the bloodied dog, its side ripped open, let out a piercing squeal, burying its head in the ground.[97]

The delay caused by the intervention of the unfortunate interloper is crucial: it allows Nikolai's old borzoi Karai to overtake the wolf and to grab him by the throat. This leads to Nikolai's moment of ecstasy as he looks down on the writhing wolf amid the pack of borzois in the passage with which I opened this chapter. While I have quoted it only in part, Tolstoy's precise portrayal of the ways in which the various dogs pursue the wolf reflects a nuanced and accurate understanding of the ways in which such encounters played out as described in other nineteenth-century sources. Machevarianov, for example, emphasizes the difference between borzois' natural tendency to close on a hare, which he contrasts to the need to train them to attack foxes and especially wolves. He notes that both lineage and practice, as well as experience alongside more seasoned dogs, are paramount in breeding and training borzois to be capable, like Nikolai's Karai, of taking down a large wolf. In addition, he emphasizes that the borzoi must grab the wolf by its throat (as does Karai) rather than lunging for its rear legs (as did Lyubim):

> A dog of such lineage will always take a wolf by the ear or the throat and will growl like a bulldog, but a dog that is simply vicious will either nip at the wolf's legs or its tail and will leap aside as soon as it turns.
>
> In order to set young dogs against a wolf, they must be put alongside a trusted, experienced and obedient wolfhound, while making sure that there are absolutely no cowardly whiners, neither young or old on the field; if not, the bad example (just as with people) is contagious, and they'll sooner follow it than the good one.[98]

Karai, despite his history of having once taken on a mature wolf alone and even with the aid of the other dogs, is unable to keep the wolf immobilized. As Nikolai prepares to dismount in order to stab him, the wolf unexpectedly manages to break free from the tangle of borzois and again heads for the woods that offer escape. It appears that the relatively inexperienced Nikolai is either insufficiently daring or simply responds to the rapid sequence of events too slowly. At this point Danilo, quicker to act and more intrepid than any of the gentry hunters, demonstrates his worth as the Rostovs' chief huntsman:

> But when the hunters did not dismount, when the wolf shook himself and again began to make off, Danilo sent his brown horse not towards the wolf, but in a straight line towards the timber, just as Karai had done, to head off the wolf... Nikolai did not see or hear Danilo until the brown horse, breathing heavily, snorted past him, and he heard the sound of a body falling and saw that Danilo was already lying on the wolf's rump in the midst of the dogs, trying to catch him by the ears. It was obvious to the hunters, and to the dogs, and to the wolf himself that it was all over now. The beast, his ears laid back fearfully, tried to get up, but the dogs clung to him. Danilo, standing up, made a falling step and with his whole weight, as if lying down to rest, collapsed on the wolf, seizing him by the ears. Nikolai wanted to stab him, but Danilo whispered, "don't, we'll truss him up," and, changing position, he placed his foot on the wolf's neck. They put a stick in the wolf's mouth, tied it with a leash like a bridle, bound his legs, and Danilo rolled the wolf from side to side a couple of times.[99]

Tolstoy's portrayal of the taking of a live wolf echoes Volunin's mythic depiction of Prince Boriatinsky. It is the Rostovs' huntsman Danilo, however, rather than Nikolai himself who ultimately captures the wolf and binds it on his master's behalf as the young count defers to his serf's whispered directives. Tolstoy explicitly parallels Danilo to Karai, noting that they share the same instinctual insight into the wolf's escape route and that Nikolai looks down on both from his horse in similar ways. It is they who physically confront and engage the wolf, mediating his confrontation with this ultimate embodiment of wild Russian nature, while he plays the part of an active observer. At the climactic moment of the hunt Nikolai proves unable to live up completely to the archetype of the nobleman wolf hunter exemplified in Prince Boriatinsky. In my view this does not represent a characterological failure on Nikolai's part but rather a realistic and convincing recognition by Tolstoy that this act might be beyond the capacity of a relatively young hunter encountering a mature wolf for the first time.

Afterwards, Tolstoy offers a final glimpse of the wolf, now trussed and tied to the saddle of a "shying, snorting horse." In addition to the old wolf, two young wolves have been taken by the scent hounds and three by other borzois:

> The hunters came together with their quarry and their stories, and everybody went to look at the seasoned old wolf, who, lolling his big-browed head with the stick gripped in his mouth, looked with wide, glassy eyes at this whole crowd of dogs and people surrounding him. When touched, he jerked his bound legs and looked at them all wildly and at the same time simply.[100]

Like Volunin's 1842 narrative, Tolstoy's account in *War and Peace* explicitly foregrounds the moment of despairing submission in which the captive wolf lies prone and bound after his violent encounter with the borzois, able only to stare mutely at his human conquerors as they touch him. Both narratives emphasize the savage otherness of the captive wolves, symbolically compressed into their opaque stares. Yet, unlike Volunin's account, Tolstoy's carries an underlying ambivalence. The wolf, trussed to the shying horse, embodies wildness that has been constrained, but not tamed; vulnerable and uncomprehending, he evokes a sense of mortality and fear. His eyes are both wild and simple, unlike the probing and intrusive eyes of the people who surround and handle him. The reader, remembering that the wolf recoiled physically at the sense of Nikolai's gaze on him earlier in the scene, can imagine the trauma the bound creature feels at the sensation of so many human eyes on him, rendering his own eyes wide and glassy with shock.

Driansky presents several examples of borzoi wolf hunts that shed additional light on the realism and accuracy of Tolstoy's depiction, as well as his relative restraint in describing the hunt's denouement. Driansky's depictions are more gruesome as he presents the encounters between borzois and wolf, as well as the hunters' use of their daggers, in more visceral detail. In each instance the wolves are killed with a dagger thrust, rather than taken alive as in Tolstoy's depiction. In one case the narrator's hunting mentor, Count Aleev, who is an experienced borzoi hunter, kills a mature wolf himself. As his borzoi, Striker (*Porazhai*), grapples alone with a wolf—"something rarely seen during a hunt"—Aleev shields himself from its jaws by approaching from its rear quarters and thrusts his dagger deep into its lower abdomen while Driansky's narrator marvels at both the count's and his borzoi's bravery.[101]

Harnessing the Domestic to Confront the Wild • CHAPTER 1 | 27

Figure 5 Nikolai Yegorovich Sverchkov, Wolf Hunting, 1870. Heritage Image Partnership Ltd/Alamy Stock Photo.

In Driansky's most detailed portrayal of a wolf hunt one of the borzois is wounded by a large wolf, as in Tolstoy's novel, leaving the others afraid to attack further with the exception of the most seasoned borzoi, who belongs to a Count Atukaev. The count arrives as the remaining dogs pile up on the wolf. As in Tolstoy's account, the count does not himself jump onto the wolf but orders one of his hunters, Egorka, to administer the coup de grâce while the wolf is held immobile by Atukaev's borzoi, Chaus:

> The Count directed them to take the beast.
>
> The hunters jumped from their horses, with Egorka in front. Grabbing the wolf by its hind leg he thrust his dagger into its groin up to its hilt. The [other] dogs jumped away, and only Chaus alone remained on the ground. His jaws clamped onto the wolf's throat and fastened in place there. The beast, wheezing, lay sprawled out. An assistant huntsman ran up to Chaus and pried open his jaws with a dagger.

> The brave fighter walked quietly off to the side amidst general praise and again fell to the earth, breathing heavily. A foam of bloody froth poured from his throat. His bloodshot eyes glittered like red-hot coals.
>
> With a delighted expression Egorka began to fasten the wolf to his saddle as a trophy that belonged to him according to the rules of the hunt.[102]

Driansky's portrayals are considerably more brutal than Tolstoy's but the basic contours are the same. Borzoi and serf huntsman grapple directly with the wolf as the gentry look on from astride their horses or, as in Volunin's retelling of Prince Boriatinsky's hunt, participate directly. In Driansky's depictions the wolves are killed immediately and pitilessly in a way that allows the hunters to protect themselves from their rending jaws, whereas in Tolstoy's novel Danilo shows his consummate mastery as a hunter by taking the wolf alive, exposing himself to the risk of being bitten in the process. One might interpret Tolstoy's relative restraint in depicting such gory and visceral killings as part of his project in writing a novel with more universal appeal, rather than one that—like Driansky's—was intended especially for an audience of hunters. In addition, his choice to have Danilo capture the wolf alive accords with the mythical underpinnings of borzoi wolf hunting in its purest form, as the beast is literally subjugated rather than killed, which provides the opportunity to depict the captive wolf's demeanor. Finally, as I will explore in depth in Chapter 4, wolves that had been captured alive were often utilized in public or private wolfhounding competitions, which hunters promoted as a way of training borzois to attack wolves and also valued as a way to show off their best borzoi wolfhounds in competition with one another.

In his guide Reutt addresses the fact that a wolf can either be taken alive from under the borzois or killed with a dagger thrust. He emphasizes the difficulty and danger of the first approach for the hunter, and also suggests that a large wolf should be killed as quickly as possible if young borzois are present to avoid frightening them. He urges his readers not to thrust a dagger into the wolf's abdominal area (as twice occurs in Driansky's portrayals), but instead to insert it between the ribs into the breast on its left side, assuming the borzois have hold of the wolf by its throat. This clearly reflects a desire to avoid further endangering the canine and human members of the hunting party rather than compassion.[103] Reutt particularly emphasizes the difficulty of binding a wolf alive:

> Not all are gifted with the art of binding. One must have the utmost courage and confidence in the strength of one's dogs to lay an unarmed

hand upon a wolf. Strictly speaking, two hunters should carry out the binding: one ought to grab the wolf by its ears while the other does the rest. Nevertheless, in testament to our hunters, one must say that in Russia, sometimes, a single person binds even an old wolf.[104]

Reutt's testimony bears a striking resemblance to Machevarianov's praise of those rare Russian hunters who can singlehandedly tie up a mature wolf, which served as this chapter's epigraph. Both men single out this act as emblematic of a national identity that depends both on the existence in Russia of such a fearful predator and of extraordinary men who can single-handedly grapple with and subdue this emblem of savage nature, although—crucially— only with the help of their bravest and most capable borzois. This context helps to underscore the magnitude of Danilo's act on Nikolai's behalf, as he so boldly exemplifies the mastery to which Nikolai aspires.

In his several articles on foxhunting informed by his perspectives as a participant-observer, Garry Marvin emphasizes that English foxhounds cannot be understood simply as flesh-and-blood animals. Rather, they are "representations" of human desire, will, and agency who exist to fulfill specific roles in an elaborate form of cultural engagement with the natural world: the foxhunt. Individual foxhounds are meant to serve as ideal exemplars of a breed, a sculpting of nature for human purposes, and a complete pack should consist of hounds that together create an aesthetically harmonious and well-functioning whole.[105] Similarly, but in a very different vein, the fox they pursue represents a set of symbolic associations that imbue the hunt with meaning—associations that have evolved over time in conformity with changing social and cultural conditions. As the fox was regarded as vermin rather than legitimate game prior to the eighteenth century, aristocratic English hunters had to "reinscribe and reimagine it" as quarry worthy of noble pursuit as they had formerly pursued boar, wolves, and bears prior to their extinction in England.[106]

Marvin's perspectives provide a useful frame of reference for considering the very different situation of borzoi wolfhounds and wolf hunting in Russia as explored in this chapter. As I've shown, Russians considered the encounter between wolf and borzoi to be a quintessential expression of the Russian empire's confrontation with wild nature as embodied in its most emblematic predator. For its part, the borzoi was the most wolflike of all hunting dogs. The very qualities of savagery, speed, and ferocity so often ascribed to wolves were those attributed to the most prized and valued borzois, and their names often mirrored these same shared qualities. Those who hunted wolves with borzois—whether gentry hunters or the serfs or others who served them as huntsmen—gained the respect of other participants in the hunt according to the degree to which they participated directly and capably in the struggle between borzoi and wolf, the domesticated canine and its untamed foe. The ultimate feat, and one which could earn legendary status for its perpetrator, was to grapple physically with the wolf alongside one's borzois. The narratives and historical context we've explored demonstrate how few were capable of this, and how much respect they garnered.

Prior to closing this chapter, I will briefly juxtapose Nikolai's experience of the borzoi wolf hunt with his subsequent conduct in the closely related semiotic sphere of warfare during a battle that takes place two years later in narrative time. This juxtaposition sheds light on several issues including his lack of compassion for the wolf, the significance of the gaze, and the role that violence (and restraint) plays in Nikolai's process of maturation. Crucially, the battle scene shows that as Nikolai grows older his personal sphere of moral consideration grows to encompass the enemy soldiers he fights, while excluding the animals he continues to hunt. This comparison will help to set the stage for further contemplation of the ways in which some Russians, including Tolstoy himself, began to extend their notions of empathy and moral concern to include wolves in Chapter 4 and the conclusion.

In 1812, during a skirmish with the invading French army, Nikolai—now in command of his own squadron of hussars—leads a daring charge to save a retreating battalion of Russian uhlans from the French attackers. From the beginning of the scene Tolstoy emphasizes the parallels with the hunt, as Nikolai appraises the field of battle. Yet crucial differences between the earlier wolf hunt and violence directed at other human beings emerge as the scene unfolds:

> Rostov, with his keen hunter's eye, was one of the first to see these blue French dragoons pursuing our uhlans . . . He sensed intuitively that if he were now to strike the French dragoons with his hussars, they would not hold out; but if he were to strike, it would have to be now, at this moment, otherwise it would be too late . . . With the feeling with which he raced to intercept a wolf, Rostov, giving his Don horse free rein, galloped to intercept the disordered lines of the French dragoons . . . Nikolai saw that in a few seconds he would overtake the enemy he had picked out as his target . . . A moment later Rostov's horse struck the officer's horse in the rump with its breast, almost knocking it down, and at the same moment Rostov, not knowing why himself, raised his saber and struck the Frenchman with it.

The moment he did this, all Rostov's animation suddenly vanished. The officer fell, not so much from the stroke of the sword, which only cut his arm slightly above the elbow, as from the jolt to his horse and from fear. Reining in his horse, Rostov sought his enemy with his eyes, to see whom he had vanquished. The French dragoon officer was hopping on the ground with one foot, the other being caught in the stirrup. Narrowing his eyes fearfully, as if expecting a new blow any second, he winced, glancing up at Rostov from below with an expression of terror. His face, pale and mud-spattered, fair-haired, young, with a dimple on the chin and light blue eyes, was not at all for the battlefield, not an enemy's face, but a most simple, homelike face. Before Rostov decided what to do with him, the officer cried out: "*Je me rends!*" [I surrender]. He hurriedly tried but was unable to disentangle his foot from the stirrup, and his frightened, light blue eyes were fixed on Rostov . . . [After the French officer's surrender] Rostov galloped away with the others, experiencing some unpleasant feeling which wrung his heart. Something unclear, confused, something he was unable to explain to himself, had been revealed to him in the capture of this officer and the blow he had given him . . .[107]

Throughout the encounter, Tolstoy emphasizes the role of the looks exchanged between the two men as they evaluate each other and decide what to do. The French officer's "light blue eyes" and frightened expression play a crucial role in Rostov's instinctual decision to show him mercy, as does the Frenchman's ability to verbalize his desire to surrender in a language both can understand. Neither of these attributes had characterized Nikolai's confrontation with the wolf, of course, and Rostov's perceptions of him correspondingly lacked empathy. Mulling over the event, for which he receives a St. George Cross for bravery and is promoted to command of a battalion, Rostov ponders the ambivalence and discomfort he felt in confronting the French officer: "And what harm had he done, with his dimple and his light blue eyes? But how frightened he was! He thought I'd kill him. Why should I kill him? My hand faltered. And they gave me the St. George Cross. I understand nothing, nothing!"[108]

Rostov's moral uncertainty represents a key moment in his maturation as a character. Just as the larger novel interrogates the meaning and justification of warfare, so does Tolstoy's protagonist gain a deeper understanding of the ambiguity and problematic nature of inflicting violence on his fellow humans through his encounter with the French officer, although curbing his proclivity for violent outbursts remains a challenge for him. At the end of the novel, now happily married to Princess Marya and a successful landowner whom the serfs respect and trust, he nevertheless continues to struggle to rein in his irascible temper and to restrain himself from striking serfs who have misbehaved. He also, however, remains a passionate hunter who devotes months each fall to the chase.

While Nikolai continues to hunt through the novel's end, Tolstoy himself would eventually come to view animals as worthy of moral consideration alongside humans, ultimately adopting vegetarianism alongside pacifism as central tenets of his personal identity and public persona, as we will see in Chapter 4. Although he represented an extreme case, the decades that followed the publication of *War and Peace* would witness larger changes in civic society that reflected some of the same tendencies, setting the stage for more empathetic perceptions and portrayals of wolves. None of the texts we have encountered thus far have presented wolves' physiognomies or behaviors in such a way as to parallel the French officer's admission of vulnerability and plea for compassion, nor evinced any significant indications of human sympathy for the hated predator. The vantage points I have explored in this chapter have uniformly emphasized those qualities that made wolves such formidable antagonists. This is precisely what rendered the borzoi hunters' victories so meaningful to them. Even Tolstoy's depiction, with its idiosyncratic references to the overwhelmed wolf's fear and distress, focused primarily on its strength and ferocity. In upcoming chapters we will explore other aspects of the ways in which Russians viewed wolves that will complement those we have seen so far. But the continuing cultural resonance of the confrontation between borzoi, hunter and wolf will continue to serve as a point of reference for each of these other portrayals.

CHAPTER 2

The Rise of Hunting Societies, the Professionalization of Wolf Expertise, and the Legal Sanctioning of Predator Control with Guns and Poison

> In Russia each of us in disorder and at random, without communication, cooperation, and help defends himself from predators using his own personal and accidental means and methods, which are indicated in our law by the terse term "all possible."[109]
>
> —**N. V. Turkin,**
> *Hunting Laws* (1889)

Borzoi wolf hunting was intimately interwoven with the mores and means of the Russian aristocracy and wealthy landed gentry. As the nineteenth century progressed the inherited wealth, private ownership of large tracts of land, and system of serfdom that had supported this form of hunting all diminished or were supplanted. Yet during the century's waning decades calls for more effective reduction of Russia's wolf populations—and even eradication of all wolves in the more populated regions of European Russia—gained increasing prominence as well. Advocates for wolf control recognized that the grand aristocratic wolf hunt on horseback had been largely superseded, but it was still considered one of the most effective means of killing wolves and finding equally effective approaches to replace it was problematic. In the late 1870s the so-called Wolf Problem rose to prominence in a way that reflected the characteristics of a new era. The emancipation of the serfs in 1861, the bureaucratic reforms that followed, the burgeoning of civil society through organizations and associations such as hunting societies, the rise of the natural sciences including zoology, the proliferation of both the specialized and popular press, the creation and expansion of a network of railways between the two capitals and out into the provinces—all these aspects of modernity influenced Russians' consideration of the Wolf Problem, which continued to be debated up to and beyond the promulgation of the epochal hunting law of February 3, 1892.

This chapter's epigraph, drawn from the work of Russia's foremost expert on hunting law, demonstrates the extent to which efforts to control predators and especially wolves became intertwined with fundamental concerns among the ruling and intellectual classes about the dissonances and disruptions that followed emancipation. While a centralized and increasingly professionalized bureaucracy sought to provide overall direction for the empire through enacting policies and laws based in statistical and scientific approaches, and newly established institutions like the elected rural *zemstvo* councils attempted to facilitate their effective implementation, much of the countryside came to suffer from what Ekaterina Pravilova has called a "vacuum of power" as the tsarist government found that it "lacked the administrative resources to create a new system of governance" to replace administration by landlords and as the population attempted to adjust to post-emancipation conditions.[110] Concern over the empire's inability to effectively control its wolves, which were perceived to have proliferated in the decades immediately following emancipation, became interwoven with deep insecurities about Russia's capabilities and prospects in comparison with those of Western Europe and trepidation over the country's ability to transcend its past. Experts compared the destruction that wolves wreaked on Russia's rural economy with scourges ranging from the cholera epidemics that ravaged its population to the fires that regularly destroyed the wooden buildings of its villages. Wolves served as a ready emblem for these overriding concerns about Russian backwardness, the latent potential for chaos, and the tensions between the promise of reforms and the difficulty of enacting them successfully.

Assertions of Russia's Wolf Problem reflected a new statistically-based and scientifically-grounded understanding of the significant effects that wolves had on the country's livestock and rural populations along with a conviction that changing social, economic, and environmental conditions in the wake of emancipation had led to an increase in the empire's wolf populations. Government researchers, provincial hunters, and zoologists debated the extent and causes of wolf depredation and proposed various solutions. During the last three decades of the nineteenth century, as sentiment among hunters and government policy were converging toward the law of 1892, which reaffirmed and institutionalized the goal of predator eradication, Russia's hunting societies came to play an especially prominent role in consideration of these issues and in the implementation of measures to reduce wolf populations. Foremost among these were the Moscow Hunting Society, established in 1862, and the Imperial Society for the Promotion of Game and Wildlife of Economic Significance and Proper Hunting, established a decade later in 1872. The two societies were fundamentally different in their orientations and priorities, and they approached Russia's Wolf Problem in divergent ways. The Moscow Society focused on providing its exclusive paying membership with satisfying experiences in the field, primarily through hunting with firearms; organized wolf hunts represented one of its most sought-after activities. The Imperial Society was more academically oriented and had farther reaching goals than the Moscow Hunting Society and it published a significant monthly journal, *Nature and Hunting* (*Priroda i okhota*), along with a related newspaper, *Hunting News* (*Okhotnich'ia gazeta*). It promoted the use of strychnine, also employed in Western Europe and North America, as an effective means of reducing wolf populations, espousing utilitarianism over traditional notions of hunting as a pastime for a privileged elite.

In this chapter I will first provide an overview of these two societies and some of their wolf-oriented activities, then focus on the Wolf Problem as it was debated primarily in the pages of *Nature and Hunting* in the 1870s and early 1880s. This will lead into a discussion of the work and significance of two of the Imperial Society for the Promotion of Game and Wildlife of Economic Significance and Proper Hunting's most accomplished and prominent figures, L. P. Sabaneev and N. V. Turkin. Sabaneev, who edited *Nature and Hunting* from 1878 to 1892, authored a lengthy 1880 monograph on wolves that combined a zoologist's perspectives with utilitarian advice about hunting techniques, as well as observations about peasants' attitudes toward wolves. His monograph provides insight into the ways in which hunters of his day perceived wolves as adversaries and the range of attitudes toward wolves among Russia's rural population. Turkin, who succeeded Sabaneev as editor of *Nature and Hunting*, developed into the country's foremost expert on Russian and international hunting law and served as a leading architect of the 1892 hunting law. He authored three substantial books on Russian hunting law over almost a quarter-century span, the earliest in 1889 and the last in 1913, as well as a book-length history of the Imperial Society's first twenty-five years in 1898. Together and in context, Sabaneev's and Turkin's contributions illustrate the ways in which debates over Russia's wolves shed light on the shifting relationships between the rural peasantry, the gentry, and the increasingly significant professional classes (including scientists, government bureaucrats, and journalists). They also indicate the importance of hunting societies as one example of burgeoning civic society, as well as the methods by which Russia tried with mixed success to emulate Western European precedent in enlisting its population to control predators. At the chapter's end I will focus briefly on a story written by the prose writer and dramatist I. A. Salov entitled "Wolves." The story describes a wolf hunt sponsored by a local hunting society in the provincial city of Saratov, located in southern Russia, that took place in the late 1800s. In keeping with my approach through the book, this literary portrayal will help to add detail and nuance to the more historical sources to which the chapter is primarily oriented.

The Moscow Hunting Society was established by a small and elite group of gentry and aristocratic hunters, including major landowners, who were concerned both with maintaining the quality of their own hunting experiences and with addressing a broader decrease in game in the Moscow region, which reflected illegal hunting and springtime poaching, especially of game birds intended for the market.[111] The Society received approval of its initial charter from the Ministry of the Interior in November 1862.[112] Although it was oriented primarily toward gun hunting and recognized the impracticality for most of its members of maintaining large kennels of dogs, the Moscow Hunting Society was an exclusive group as evidenced by one of its earliest and most famous hunts, which took place in the province of Vladimir on December 10, 1862 with Tsar Alexander II as the guest of honor. After their huntsmen had located a bear within thirty miles of a railway line in the Pokrovsky district and a special committee had renovated the village of Golovino during the week before his arrival, twenty-six members of the Society joined the tsar to await the bear as it was driven toward their line by the loud shouts of their huntsmen and members of the local populace. Alexander was placed with his weapons and royal suite in the most advantageous spot to intercept the bear, flanked by Society members to the left and right, and was eventually able to kill it with a succession of shots.[113]

According to its compiled membership lists, the Moscow Hunting Society included nine members of the royal family as honorary members, along with a long list of titled nobility and other members of the gentry among those who paid dues (although it was ostensibly open to candidates from all classes via the normal election procedure of a two-thirds vote of members).[114] Its active membership ranged between 100 and 200 from the mid-1860s until 1913, peaking at 188 in 1885.[115] At its inception in 1862 its yearly dues for paying members were thirty rubles.[116] By 1913 they had increased to 150 rubles per year, another indication of its exclusive nature.[117] These fees entitled members to make use of the Society's extensive rented forests, marshes, and meadowlands for bird and small game hunting as well to participate in organized hunts for wolves, bear, foxes and other predators, which typically took place in the fall and winter. It leased lands primarily within a 100-mile radius of Moscow although on occasion members traveled twice that distance or more to hunt, aided by the development of the railway system, and in 1884 it prevailed on the imperial administration to grant it exclusive rights to hunt for wolves on certain state lands in the provinces of Moscow and Vladimir.[118]

The Moscow Hunting Society espoused predator control as one of its priorities and developed a special reputation for hunting wolves according to a drive technique pioneered in the Pskov region. This form of organized hunting gained currency as a method that could allow gun hunters to achieve results rivaling those attained by men on horseback with large packs of scent hounds and borzois. Starting in the winter of 1867-68, the Society kept under contract between three and seven hunters trained in the Pskov method, so-called *pskovichi*.[119] They would pinpoint a den site or the current location of a pack of wolves, then plan a drive so that the lay of the land and obstacles blocking escape avenues would drive the wolves toward the waiting hunters. Members, who paid an additional fee of two rubles to participate in these winter hunts, were stationed by number in predetermined positions decided by the casting of lots and were restricted to shooting wolves that entered their areas.[120] Combined with a new generation of cartridge-fired guns that became widely available in the early 1880s, these drives enabled the Society to embark on a campaign of exterminating local wolf populations, and by the mid-1890s wolf numbers had been substantially reduced in the province of Moscow. In March 1895 Grand Duke Sergei Alexandrovich took the Society under his protection, specifically citing its membership's contribution to the wellbeing of the peasant population through its wolf hunts.[121] On December 6, 1897 the 1000th wolf was killed on a Society-sponsored hunt, and in 1912 the 2000th, already much farther afield near the Chiplyaeva station on the Riazan-Ural railway line.[122]

While the Moscow Hunting Society's preeminent concern was to provide such satisfying experiences in the field to its members, rather than to systematically eliminate wolf populations throughout European Russia, the Imperial Society for the Promotion of Game and Wildlife of Economic Significance and Proper Hunting had greater ambitions and farther-reaching impacts.[123] Established as an offshoot of the Imperial Moscow Society of Naturalists in 1872, it actively encouraged its members to contribute to the developing fields of natural history and zoology as field researchers.[124] It benefited from the direct sponsorship of Grand Duke Vladimir Alexandrovich and was the wealthiest of Russia's hunting societies, as it did not rely purely on membership dues. The Imperial Society viewed its mandate as evaluating the significance of sport, subsistence, and commercial hunting throughout Imperial Russia—though in practice it focused primarily on the more populated provinces west of the Urals—as well as promoting practices that would help to secure the future of hunting, trapping, and animal husbandry across the empire's territories. Turkin summarized its goals in his exhaustive 1898 history of the Imperial Society's first quarter century:

> The tasks confronted by the Imperial Society, as already mentioned, were extremely broad and multifaceted. The most urgent to be carried out was a comprehensive analysis of hunting and associated trades in Russia, and directing them upon a more rational course; the preservation of useful animals and the extermination of harmful ones; the discovery of means to increase the reproduction and breeding of creatures that serve as the object of hunting and other trades; the establishment of a museum, a library, and a hospital to promote proper dog breeding in Russia through exhibitions; and, finally, publication of a specialized hunting journal.[125]

From July 1874 through the end of 1877 the Imperial Society published the *Journal of Hunting* under the editorship of L. P. Sabaneev (1844-1898). Sabaneev, who had been trained in zoology and the natural sciences at Moscow University, served as secretary of the Imperial Moscow Society of Naturalists (MOIP) during the 1860s and subsequently edited the journal *Nature*, founded in 1873.[126] After three years a decision was made to combine the two journals under Sabaneev's editorship. The resulting monthly journal, *Nature and Hunting*, was intended to appeal to a wider audience than either of the more specialized publications. Edited by Sabaneev through the early 1890s and subsequently by Turkin, it became Imperial Russia's most significant hunting periodical, appearing monthly from 1878 to 1912.[127] This was an impressive accomplishment that attested to the Imperial Society's significance as well as the energy and resources of the journal's two successive editors. During this period the range of articles and other materials that it featured were extraordinarily diverse, as I explained in the introduction. This breadth allowed *Nature and Hunting* to attract a broad array of contributors and to appeal to a variety of readers, expanding its reach beyond the community of dedicated hunters.

As part of its overall effort to collect information and establish links with remote corners of the empire, the Imperial Society also encouraged the establishment of provincial branches. From 1874 through the turn of the century a number of these came into existence so that by 1897 the Society had branches in fifteen provinces, representing nearly half the territory of European Russia. Attempts to establish branches in the Empire's northern and eastern territories were less successful, whereas they were established as far west and south as Warsaw (1888), Tiflis (1891), Odessa (1894), and Riga (1897). The northernmost branch in Russia proper was established in 1875 in Vologda, a provincial capital 300 miles northeast of Moscow. A St. Petersburg branch was established in 1880. Turkin documented abortive attempts to establish outposts in Ekaterinburg just east of the Ural Mountains (1875) and in Irkutsk on Lake Baikal (1897), as well as in the Zabaikal area near the Chinese border (1888), but he provided little information about these in his history.[128]

From the mid-1870s on, using its journal as a clearing house of information and drawing on its growing network of provincial branches, the Imperial Society set about gathering information on predator and prey populations and the hunting economy throughout the empire. It solicited reports from provincial governors and other administrators, as well as ordinary hunters and citizens. As Turkin noted: "This was the first overall investigation of Russia's hunting economy, and from this time onward the basis for systematic yearly collection of information about hunting was established in certain provinces."[129] The damage inflicted by wolves and other predators was a consistent theme in many of these reports. For example, the figures submitted by the administration of the northern Vologodsky Province were broken down for 1872 to

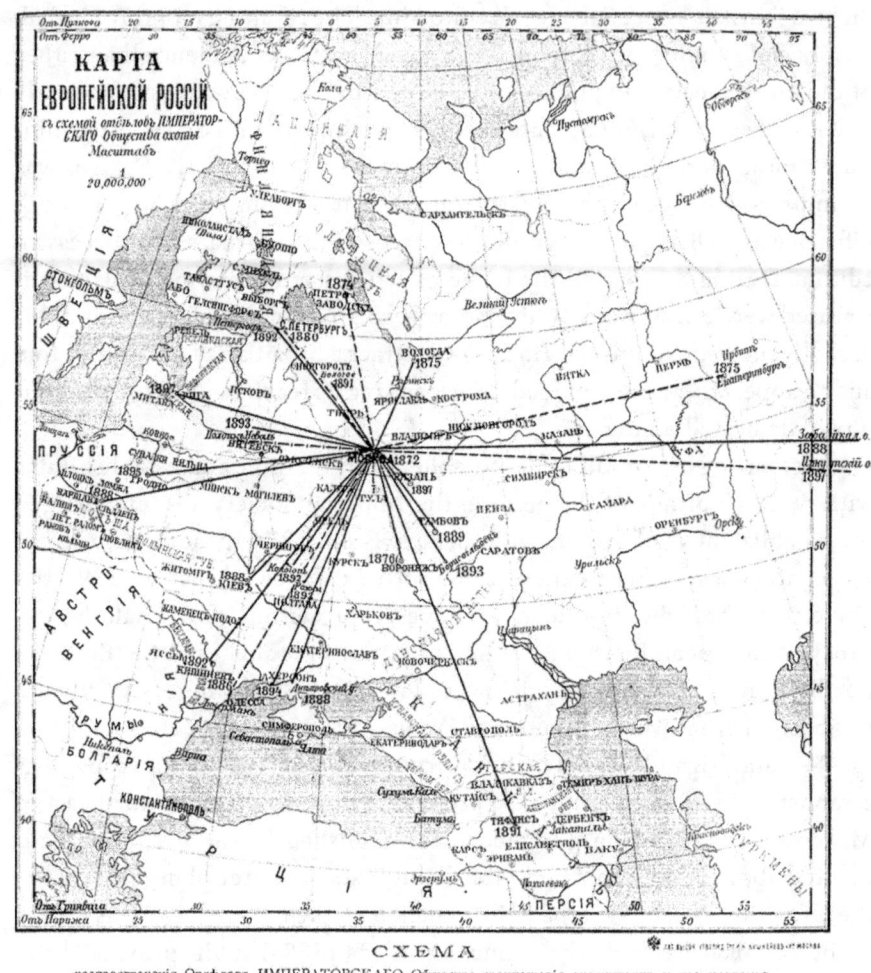

Figure 6 Branches of the Imperial Society for the Promotion of Game and Wildlife of Economic Significance and Proper Hunting as of 1897. From N. V. Turkin, "Istoricheskii ocherk deiatel'nosti Imperatorskago Obshchestva razmnozheniia okhotnich'ikh i promyslovykh zhivotnykh i pravil'noi okhoty za 25-letnii period sushchestvovaniia, 1872-1897," *Priroda i okhota*, January 1898.

1874 by district and by categories of large and small livestock. In certain of the province's districts, according to these figures, economic losses caused primarily by wolf predation amounted to more than 40,000 rubles a year, and the economic losses to the province as a whole surpassed 150,000 rubles in a typical year.¹³⁰ Summarizing the information gleaned from a number of the provincial reports that were submitted in the mid-1870s, Turkin stated:

> It became clear that even in the most far-flung corners of the empire, where civilization had not yet penetrated, the population of commercially valuable species and those of interest to hunters had appreciably decreased, and that the main reasons for this were universally ruinous means of practicing hunting and associated trades and rising populations of predatory animals that were destroying useful game. Crowning this was the widespread discontinuance of hunting with borzois, the absence of which had inflicted a significant toll upon the rural population.¹³¹

As this overview indicates Russians of this era, including experts like Turkin, were convinced both that the number of wolves had increased after the emancipation—which they attributed partly to the decline of borzoi wolf hunting—and that wolves were taking a substantial toll on Russia's livestock and rural agricultural communities. They diverged, however, in their varying appraisals of the nature and extent of the threat and the best approaches to dealing with it. During the 1870s concern over the Wolf Problem grew to a crescendo, which would influence the activities of Russia's hunting societies and the very contours of the empire's hunting laws. A pivotal publication in the debate over Russia's Wolf Problem, entitled *On the Destruction of Domestic Livestock and Wild Game by Wolves and on the Eradication of Wolves,* appeared in 1876. Its author, V. M. Lazarevsky (1817-1890), had risen through a long bureaucratic career during which he also published literary works and aided V. I. Dal' in compiling his famous dictionary of the Russian language to the position of a Vice-Director in the Ministry of the Interior and appointment as a Privy Counselor in 1873.¹³² Lazarevsky was commissioned by the tsarist government to investigate the Wolf Problem and his findings, based on extensive analysis of statistics from forty-five provinces located west of the Ural Mountains for the year 1873, were published as a supplement to the *Government Bulletin.*¹³³

Lazarevsky's statistical analysis led him to conclude that across European Russia 179,000 large and 562,900 smaller domestic animals had been killed by wolves in 1873. The former category included horses and cattle, while the latter included the young of these species and smaller domestic animals. He equated this with a financial loss to the country and its inhabitants of seven-and-a-half million rubles. He surmised that the actual loss of livestock might well be twice as much as the official figures indicated, corresponding to a loss instead of 15 million rubles across the forty-five provinces, not including the value of dogs or domestic fowl taken by wolves.[134] He then speculated further: if one took into consideration a wolf's hypothetical daily need for meat (which he considered to be 7 pounds), estimated a wolf population of 180,000 to 200,000 west of the Urals, and assumed that any meat not obtained from domestic animals must come from wild game at prevailing market values, one could estimate the value of all wildlife lost to wolves in European Russia at 50 million rubles.[135] In addition to these economic impacts, he noted, wolves killed a portion of Russia's rural population each year. From 1849 to 1851, for example, 266 adults and 110 children had been killed by wolves, or an average of about 125 people per year, as reported in the Journal of the Ministry of Internal Affairs.[136] Lazarevsky compared these losses of livestock, property and life to those that resulted from outbreaks of cattle plague and fires, concluding that wolves caused great harm to the empire's rural economy than either of these other afflictions.[137]

Having demonstrated to his statistical satisfaction the dire effects of wolves on Russia's rural economy, Lazarevsky then critiqued the range of methods typically employed to hunt wolves. He singled out Russia's local authorities and rural elected *zemstvo* councils for not addressing the problem with commensurate seriousness and resources—for example by providing sufficient bounties for wolf pelts—and enumerated in detail twenty examples of inadequate administrative oversight and coordination drawn from perusal of *zemstvo* bulletins to prove his point.[138] He concluded:

> The indifference is truly shocking and can be explained only by the ignorance of the *zemstvo* about what is happening in its local economy. It's true also that incidents of wolf depredations are spread out, that they don't hit you in the eyes, and that they've become used to this misfortune as if it were one of the conditions of our normal agricultural reality.[139]

Lazarevsky concluded that the only practical means of eradicating wolves lay in poisoning them with strychnine, a method that had been employed in West European countries like France since at least the 1820s but that had been used only sporadically in Russia. At the end of his brochure, he appended instructions for preparing strychnine pills according to a method pioneered by the apothecary and hunter O. E. Valevsky and advice on utilizing the poison against wolves.[140]

In addition to his statistical analysis, Lazarevsky devoted a number of pages to describing wolf behavior, drawing on his own experiences as a hunter as well as both foreign and domestic accounts. He particularly emphasized wolves' intelligence in a passage that Sabaneev would later critique, as we shall see:

> The wolf is generally a great master of everyday affairs and doesn't act without calculation. If an animal doesn't succumb to individual attack for one reason or another, he'll form a group that cooperates. In this situation, with a complex organization of the task, each understands to perfection the role for which he is jointly responsible: one decoys the victim, another distracts its gaze, a third heads for the flank or falls back to cut off its retreat, etc. For each hunt he has a special approach.[141]

Lazarevsky's brochure, which followed decades of speculation about the value of livestock and wildlife lost to wolves, which so forcefully vilified them, and which coincided with the Imperial Society for the Promotion of Game and Wildlife of Economic Significance and Proper Hunting's efforts to assess the effects of predation on the empire's game populations, provided momentum for a renewed discussion of the desirability and means of eradicating wolves among Russia's hunters. The issue was debated with particular intensity in the *Journal of Hunting* and its successor *Nature and Hunting* as both experts and amateur hunters weighed in on the issue.

Sabaneev himself was one of the first to provide a thoughtful appraisal of Lazarevsky's brochure. In an 1876 article in the *Journal of Hunting* Sabaneev first critiqued aspects of Lazarevsky's statistical approach, concluding that his brochure appeared to substantially inflate the value of losses caused by wolves.[142] He then posed the question of whether wolves could indeed be eradicated in Russia, as they had been in countries like England. Like Lazarevsky, he maintained that poisoning wolves with strychnine was the most efficient and promising means of reducing or eliminating wolf populations. He explicitly defended the use of strychnine by noting that it was legal, that concerns about the accidental poisoning of hunting dogs and domestic animals were overblown, and that it had been widely used throughout Europe in relatively successful wolf reduction campaigns. Citing a case in which local authorities in the Russian countryside had objected to the poisoning of a wolf that had bitten ten children over the course of two to three months in part because "it is a great sin to destroy one of God's creations so inhumanely," Sabaneev asserted that poisoning was the "cheapest, most convenient, and most effective means of eliminating destructive predatory beasts."[143] While it would be impossible to completely eliminate wolves throughout Russia, he admitted, Lazarevsky's inclusion of detailed instructions and a list of ingredients for the preparation of strychnine pills should surely encourage more widespread use of poison and lead to a substantial reduction in European Russia's wolf population.[144]

Two years later an 1878 article in *Nature and Hunting* by V. Belov reported on the findings of a commission appointed by the Ministry of the Interior to consider Lazarevsky's pamphlet. The commission had concluded that undertaking a concerted campaign to dramatically reduce or exterminate wolf populations was essential and that poisoning with strychnine represented the best solution "because other methods such as hunting with hounds, trapping, and drives have not led to satisfactory results up to this point and moreover drives, which require a marshaling of the population, represent a natural obligation [but] one not established by law."[145] In addition, the commission asserted that strychnine should not pose an undue danger to livestock or dogs if handled appropriately. In light of this, Belov noted, its members had resolved to facilitate the reprinting of Lazarevsky's pamphlet and its distribution to provincial authorities in advance of anticipated changes to the legal code that could have more significant impacts. He concluded by providing examples of districts that had decided to approve the use of poison as well as bounties of typically two to three rubles per wolf, along with some that were resisting the distribution of poison, feeling the matter needed further study.[146]

Over the following years, contributors to *Nature and Hunting* continued to discuss the dangers posed by wolves to livestock and people, the incidence of rabies among wolves and wild dogs, who should hunt wolves and what methods should be employed, what compensation should be offered to hunters for killing wolves, and the effects of wolf numbers on populations of prey animals. Their contributions demonstrated the considerable variation of opinion that continued to divide Russian hunters and others interested in wildlife even after Lazarevsky's brochure and its endorsement by the authorities.

An 1878 submission from a hunter in the Yamburg District in the western portion of the St. Petersburg Province, for example, noted that wolf populations varied greatly from year to year. Even in years when they were high and caused problems, however, local peasants tended to avoid hunting them. Although the local *zemstvo* council offered bounties of six or seven rubles, not every peasant would risk confronting a mature wolf for this amount. In addition, the author claimed that many peasants believed that adult wolves would take revenge on the surrounding villages if their pups were harmed. Those who did pursue wolves tended to use either traps or poison, and in the latter case they just as often ended up poisoning dogs, whose owners therefore opposed the practice. In March 1886, for example, eighteen local dogs, including two valuable hunting dogs, had died of poisoning while no wolves had succumbed.[147]

A January 1880 submission to *Nature and Hunting* emphasized some of the same points, although without the same reservations about the use of poison. It asserted that the wolf population was increasing and causing serious livestock losses in the northeastern provinces of Russia, then called for "radical measures" to combat them.[148] It advocated hunting with dogs and poisoning as among the most effective methods, and argued that local *zemstvo* councils should provide generous bounties to subsidize hunting clubs in their pursuits of wolves with both dogs and guns, as well as poisoning, perhaps imposing a tax on livestock to provide a source of funds.[149]

Figure 7 Hunting wolves in Russia, 1860. *Illustrated Times.* Antiqua Print Gallery/Alamy Stock Photo

In the same issue of the journal another contributor, in contrast, asserted that not all hunters agreed that wolf populations were excessive. He objected especially vehemently to calls for enlisting the army in wolf eradication programs. Citing his thirty years of experience as a wolf hunter, he noted that wolves had become harder to find and argued for hunting them "properly" with scent hounds and borzois:

> In my opinion a genuine borzoi hunter enjoys hunting wolves, foxes, and hares equally. At the same time, of course, hunting that is difficult and therefore requires great skill and knowledge is the most interesting sort for him, and the complete destruction of any animal that can be hunted with borzois would be a sad thing.[150]

As these examples illustrate, Russian hunters disagreed about fundamental issues relating to wolves. Had their populations indeed increased after the emancipation, and if so was this the case everywhere? Were peasants either naïve or superstitious (or both) in their attitudes toward wolves? Which methods of hunting, trapping or poisoning could most effectively be utilized to control wolf numbers in the post-emancipation era, and how did these vary in different parts of the empire? What role should gentry hunters, hunting societies, and local government play in wolf control? Could borzois still play a significant part in hunting wolves? And finally, what characteristics did wolves possess that made them so uniquely problematic?

In the context of these ongoing debates, Sabaneev emerged as Russia's leading expert on wolves, albeit one who embodied the strongly anti-wolf stances of the day. Starting with the appraisal of Lazarevsky's 1876 brochure summarized earlier, he published a series of articles on wolves in the *Journal of Hunting* and *Nature and Hunting* over the following years. In 1880 Sabaneev organized and compiled these into a 200-page monograph published as a supplement to *Nature and Hunting* entitled *The Wolf*.[151] The monograph was divided into an introduction, a nearly 100-page natural history (*estestvennaya istoriia*) of the wolf, and an even longer section devoted to means of killing wolves, which was itself divided into subsections on hunting with dogs, gun hunting, and non-sport or professional methods including trapping and poisoning. Sabaneev's treatise represented the most significant study of wolves in Russia up to that time and reflected the significant developments that academic zoology had made in Russia during the preceding decades.

In a historical overview of Russian zoology from Peter I's founding of the Academy of Sciences in St. Petersburg in 1724 to the late Soviet period, V. S. Shishkin divides its progression in imperial Russia into three phases: an initial period focused on exploratory expeditions and classification of the empire's fauna (early eighteenth to the early nineteenth century); a second period during which the natural and zoological sciences became integrated into the curricula of the empire's recently established universities and the first scientific societies began to form (early nineteenth century to the reform era of the early 1860s); and a third period during which democratization and scientific breakthroughs such as Darwin's theory of evolution led to significant advances that affected both university curricula and society at large through the proliferation of scientific societies and journals intended for a wider readership (mid-nineteenth century to end of the Imperial period).[152] Sabaneev, of noble background but trained to candidate status in the natural sciences at Moscow University, the editor of journals that appealed to both specialists and the public, and the author of popular works on topics ranging from the fauna of the Ural region to breeds of hunting dogs, served as an ideal representative of this third era.

Sabaneev's eight-page introduction linked Russia's prevailing problems with wolves directly to aspects of the post-emancipation order in a miniature socioeconomic analysis that revealed his concern for the welfare of Russia's rural peasantry in the wake of the emancipation. While reiterating his conviction that Lazarevsky's 1876 brochure had exaggerated the total losses of livestock to wolves, he nevertheless affirmed that predation by wolves on livestock and game had increased substantially after the 1861 emancipation and was now among the most significant burdens born by peasants throughout Russia, posing even more of a threat to those engaged in subsistence agriculture than rapidly diminishing populations of game and fish:

> Over the last decade the number of wolves has increased dramatically and continues to do so with each passing year. Wolves are becoming a public pestilence, a national scourge: they run into the cities and even the capitals in broad daylight, and in the villages they approach herds and throttle livestock without fear or danger ... Our peasants, who have just recently been freed from slavery, who have scarcely ceased paying a heavy tribute to landowners, have once again fallen into servitude—only this time not to people, but rather to a predatory beast.[153]

Sabaneev claimed that the harm wolves inflicted on rural inhabitants of the empire from the northern tundra to the steppes of Central Asia, from Poland to the Amur region, were so consequential that they should be considered Russia's single most significant wild animal because of their widespread negative impacts. Therefore, he asserted, all efforts must be made to reduce their numbers, even if eradication should prove impossible: "it's long past time to recognize that the wolf is the most bitter enemy of our well-being, to throw aside our usual apathy and indifference, and to declare ruthless, implacable war on the predator."[154]

An underlying theme of Sabaneev's introduction was that Russian approaches to wolves had lagged behind the more energetic anti-wolf campaigns of Western Europe, where wolf populations had been dramatically reduced or eliminated. How, Sabaneev asked, could Russia's increasing level of civilization have resulted in an increase in wolf populations, whereas the opposite had occurred in most of Western Europe? The answer, he asserted, was that wolves had benefited from the particular stage of development at which Russia found itself: former extensive areas of unbroken forest had been partly cut down, yielding a mixed landscape of woods, fields, and marshes in which isolated villages provided easy access to domestic animals as prey, but Russian society had not yet developed socioeconomically and politically to the point of organizing and mobilizing to defend itself effectively:

> Most of Western Europe has already lived through this era but no one, surely, will deny that we are only embarking upon it and in addition are living through a transitional, and therefore, difficult period. The emancipation of the peasants, which provided us with a strong push along the path of civilization and at the same time disrupted the old structures of our lives ... was bound to bring about an increase in the number of wolves ... The professional [hunter] of the central provinces, enticed by more reliable and easier prey, and given unjust and even unlawful hindrances relating to the main weapons in the struggle with the wolf—the trap and poison—naturally has been deflected from this toilsome and unremunerative form of hunting ... In addition, the difficult period that landowners continue to live through has deprived them of the possibility of maintaining large kennels of hunting dogs, which had helped to keep the numbers of wolves in check.[155]

Sabaneev directed his critique of Russia's inadequate efforts to combat wolves across classes. He also bluntly linked the seriousness of Russia's Wolf Problem with his country's backwardness in comparison with the more advanced societies of Western Europe. He emphasized that wolves, as a parasite on livestock and rural agriculture, had seen a resurgence amid the disruptions that followed the 1861 emancipation. He noted that newly formed clubs and associations of gun hunters unwittingly provided sanctuary for wolves on their rented lands by forbidding others access in their attempts to increase game populations, although he expressed the hope that hunters' groups might eventually serve as part of the solution to the Wolf Problem:

> Without in any way denying the possibly very important future role of gun hunters in wolf eradication we, nevertheless, consider ourselves entirely justified in asserting that they, albeit indirectly and moreover unwittingly and guided by the best intentions, have often contributed to increasing populations of the beast. Borzoi hunters, when they forbade hunting on their properties, preserving the wolves for themselves, had a basis for this prohibition. Gun hunters, however, and the ever-growing hunting circles and societies—when they forbid hunting in their rented forests and marshes—become the unintentional protectors of the wolves that are born there, from whom the renters themselves have not possessed the means or desire to exact retribution . . . Consequently, those who have a moral obligation to concern themselves with the continual and unhampered destruction of predatory beasts, serve unwittingly to shelter them.[156]

Sabaneev also critiqued the peasantry, and with less caution than he utilized in criticizing his social equals. He emphasized that peasants did not regularly participate in collective efforts to kill predators as they commonly did in certain West European countries. They also habitually entrusted an entire village's herd to young or "simple-minded" shepherds, and they continued to suffer from superstitions that hindered their ability to defend themselves against wolves:

> Among our rural agricultural population there continues to exist a strange belief, bereft of any foundation, about a mother wolf's vengeance; and because of this superstition our peasants not only don't kill wolf pups, whose whereabouts are almost always well-known to them, but even hinder their destruction. This superstition, to which we will return, however, is by no means the only one. Up to this day peasants feel a superstitious fear of wolves, and ancient legends about were wolves, vampire wolves [*o volkakh-oborotniakh, volkolakakh*] still have their gullible defenders among both the Great and Little Russians. This explains why they ascribe so much intelligence to the wolf, why he is considered to be a

fated, inescapable, irresistible evil, a sort of heavenly retribution. The wolf only takes the livestock that has been determined for him by God; that meat belongs to him by right, and if you deprive him of a slaughtered cow or sheep he will take another.[157]

Sabaneev's introduction, taken as a whole, portrayed a nation in which the crumbling of the old social order of serfdom had left the population vulnerable to a dangerous predator against which new social institutions and practices were ineffective. The peasants, pursuing subsistence agriculture and gripped by superstition and misconceptions about wolf behavior—some based in a religiously-grounded fatalism that Sabaneev, as a trained zoologist, dismissed out of hand—were unable to confront the wolves who relentlessly attacked their livestock and occasionally vulnerable people, especially children. The privileged members of hunting clubs, most of them members of the gentry, were more concerned with the quality of their own game and bird hunting than with protecting their less fortunate countrymen from rising populations of wolves. And professional hunters and trappers, acting from a utilitarian calculus, found the inconsistent and inadequate bounty system, along with varying strictures on the use of poison, insufficient motivation and even a disincentive to pursue this challenging quarry.

Sabaneev's critique echoes some of the same issues that scholars such as Jane Costlow and Ekaterina Pravilova have explored in relation to Russia's Forest Question.[158] Just as the contested ownership, utilization and all too often destruction of the country's forests reflected unresolved tensions in the post-emancipation order, so did debates over the empire's wildlife, including its most fearsome predator. Achieving and implementing a consistent set of policies that could help to preserve the empire's natural resources was a continual challenge that often pitted the perspectives and needs of different social classes against each other. In addition, hunting in Russia was a more egalitarian proposition than in much of Western Europe.[159] Hunting policies for both predators and game (*dich'*) had to take into account the needs of a variety of constituencies ranging from large landowners, to the members of Russia's proliferating hunting clubs, to peasants who hunted on occasion, to market hunters and the indigenous populations of Siberia and the far east that relied on hunting and trapping for their livelihoods.

The next nearly 100 pages of Sabaneev's monograph were devoted to a historical and zoological description of wolf populations across the Russian Empire (with reference to worldwide distribution and variation) along with a discussion of wolf habits and habitats, procreation, predation, and other salient characteristics. During the course of his discussion he cited a gamut of scientific and other literature in Latin, German, French, and Russian ranging from Linnaeus to contemporary explorers of the Russian empire's hinterlands (he himself had led Moscow University trips to the Urals in the mid-1860s). He took particular pains to present a variety of Russian perspectives on wolves—including both those with which he agreed and disagreed—interpreting these accounts through personal experience and simultaneously assessing the current state of zoological knowledge of wolves in his country. He noted that wolves' current worldwide geographic distribution reflected both their habitat requirements and the policies and attitudes in place across societies and cultures to control them. These interlinked aspects determined their prevalence. He distinguished parts of Western Europe like the British Isles, where wolves had been eradicated for centuries, with those like France, where they still existed in appreciable numbers, contrasting both situations with their much greater abundance in parts of the Russian empire.

Noting that wolves had likely developed in landscapes such as the more open and mountainous regions of Central Asia and were not naturally denizens of deep and unbroken forest, Sabaneev outlined the types of habitat and prey most conducive to robust wolf populations. In all of the Russian empire, he argued, wolves were most prevalent and caused the greatest harm in eleven provinces of European Russia.[160] Central Russia's mixture of broken forests, isolated villages, scattered herds of livestock, large parcels of land off limits to hunting except by a select few, and its light population density compared with most of Western Europe, provided ideal conditions for thriving wolf populations, he argued. In the westernmost provinces adjacent to Prussia, where less land was owned by large landowners or reserved by hunting clubs and poisoning was more commonly practiced, wolf populations were appreciably lower. Wolves were also less problematic east of the Urals in Siberia, where both human and livestock populations existed in lower densities.[161]

Sabaneev devoted the next portion of his monograph to analysing wolf behavior, senses, and intelligence. Drawing in part on his personal experiences in the field, he argued that both hunters and zoologists tended to overstate the acuity of wolves' senses and especially the sharpness of their vision and sense of smell, attributing almost preternatural sensory perceptions to them (a set of misconceptions that by implication they shared with the peasantry, although presumably with less of an admixture of superstition). In fact, wolves have trouble discerning motionless hunters even in close proximity, he argued, and most scent hounds possess a more acute ability to smell.[162] Of the five senses, he argued that wolves' hearing is the most unusually sensitive.[163]

Sabaneev took particular pains to refute his contemporaries' convictions about wolves' "renowned intelligence and cunning."[164] He first took Lazarevsky to task, deriding his tendency to ascribe anthropomorphic attributes of teamwork and cooperation—and even the ability to employ strategies and ruses—to wolves and paraphrasing the passage that I cited earlier. In fact, Sabaneev argued, wolf packs are simply made up of older wolves and their progeny who fight among themselves and sometimes even kill and eat each other. Wolves' intelligence, he argued, is sufficient to enable them to attack prey from multiple directions, or outwit a village dog, but likely less than that of a fox, given that they are easier to trap or poison than the latter.[165] He asserted that wolves rely primarily on their strength, teeth, indefatigability and speed rather than innate intelligence in pursuing and killing prey. He also noted that rabies and distemper are among the only illnesses to which wolves are commonly susceptible but criticized the common belief that a mother wolf would become rabid if her young pups were taken from the den and killed.[166]

Sabaneev's vehemence in dismissing claims of wolves' superior intelligence, while directed especially at Lazarevsky, reflected his rejection of prevailing beliefs among many Russian hunters that Russia's wolves were both highly intelligent and shrewd. An 1854 article that had appeared in *The Journal of Horse Husbandry and Hunting* represented one example of the "outdated" beliefs (which he felt that Lazarevsky had helped to perpetuate) that Sabaneev was trying to overcome through his more modern zoological approach. It focused on the cunning and boldness (*khitrost' i smelost'*) of wolves and particularly those that dwell in proximity to humans.[167] It further emphasized that their hunting methods and other behaviors indicate a level of intelligence that distinguishes them from other animals and makes them into formidable foes:

> These beasts, in our climate, differ very distinctly from other wild animals in all respects: in the first place, because of their large numbers; in the second, because of their ability to carry out such tactical activities during their depredations that you sometimes find yourself at a loss when you see that their cunning overcomes all human resourcefulness and foresight. The main thing, though, that incites men against wolves is the devastation they wreak, which is always so painfully felt in rural husbandry: the destruction of sheep, horses, and cattle.[168]

The author related various anecdotes that demonstrated wolves' fiendish cleverness. He asserted, for instance, that they are far more intelligent than domestic dogs, and told of how a lone wolf would lure a dog away from its village by feigning fear or illness until the entire pack could fall on the hapless victim. Like humans, packs of wolves hunting in forests would split up so that they could lie in wait separately along the potential escape routes of hares and other prey. Such examples led him to conclude that "wolves possess a great ability to think and, apparently, to understand circumstances, which they are sometimes able to turn to their advantage."[169] Their cunning intelligence allowed wolves to fool not only village dogs and wild game but also the villagers themselves, he asserted. Some wolves—particularly those who had dens nearby and therefore wish to avoid confrontation—would hide their aggression so effectively that villagers were lulled into false complacency, claiming: "these are our kind wolves. They won't touch anyone—even if a child approaches them, they won't attack it."[170] The implicit message here, of course, was that those who knew better—i.e., gentry hunters—had to protect the trustful villagers from their own naiveté.

While Sabaneev aimed to discredit prevalent portrayals of wolves as diabolically clever, he also attempted to convey to his readers a more nuanced understanding of wolves' social life and acknowledged that they possess certain positive attributes. He emphasized the maternal and paternal instincts of adult wolves, the tendency of mature wolves to form pair bonds as part of parenting, and the particular gentleness of the female with her pups. He also presented a detailed description of wolves' lives through the four seasons focusing on their mating and denning behaviors, the role of adolescent wolves in the pack, and the various stages of wolf maturation.[171] He even went so far as to present a brief sketch of a day in the life of a wolf pack centering on a den of young pups.[172] He also emphasized that wolves tend not to kill livestock in the vicinity of their den, presumably to protect their pups from reprisal, echoing the 1854 author's conviction that this would lull local peasants into false complacency:

> Wolves spare domestic animals in the vicinity of their den because they fear revealing its location to men and initially, when there's enough sustenance available all around, livestock can forage almost right next to a wolf den. This explains why peasants of neighboring villages not only don't make an effort to kill young wolves but even strive to conceal their whereabouts.[173]

Sabaneev described in great detail the changing diet of wolves throughout the year and across different regions. In addition to livestock and especially the young of larger domestic animals during spring and summer, he noted that their diets included a variety of wild game ranging from rabbits and hares, to ground nesting birds and their eggs, to mice and other rodents, as well as many other species.[174] As for domestic livestock, wolves were opportunists and would take any vulnerable animals ranging from fowl to horses, although sheep were most often their preferred prey because of their stupidity and inability to defend themselves.[175]

Sabaneev's zoological portrayal of wolves is partially consistent with the findings of modern wildlife biology, though of course it predates the development of modern ecology with its emphasis on the interrelationships of various flora and fauna in a well-functioning ecosystem and the especially vital role of apex predators (to which I shall return). He also recognized the limitations of zoological knowledge in his day by—for example—differentiating subspecies (*podvidy*) of wolves through physical characteristics and range but acknowledging that such categorizations were provisional and should be treated as conditional, rather than absolute. His attempts to refute contemporaries' belief in wolves' intelligence and to speculate about their social behaviors are more questionable, although these areas represent vibrant subfields of modern wolf research.

In the remaining and longest portion of his monograph, Sabaneev delved in extraordinary detail into the various means by which wolves were typically hunted or otherwise killed in Russia. In this sense, his monograph echoes the borzoi hunting guides and manuals I explored in Chapter 1 updated for a more modern age. He differentiated between "sport" hunting and commercial hunting, trapping, and poisoning and between hunting with hounds versus guns, as well as less common techniques. He emphasized that commercial hunters tended to focus on more valuable and less dangerous prey, so that—in his estimation—more than half of the wolves killed in Russia each year were taken by amateur hunters with dogs or guns rather than by professionals or other non-sport hunters or trappers for bounties or the value of their pelts.[176]

He then outlined some of the fundamental differences between hunting wolves with dogs and with guns, as well as the reasons for the gradual erosion of the former practice. Scent hounds and borzois provided the most effective means of hunting wolves, he asserted. Borzois could more reliably take down a wolf at a greater distance than gun hunters; the practice of hunting with borzois was well-established with a strict set of procedures that gun hunters had not yet settled on; and borzoi hunting took place in the fall while wolves were still near their den sites, whereas gun hunting typically took place in the winter when they were more widely dispersed.[177] He argued that Russia's rural peasant populations had historically recognized the value of borzoi hunting and continued to support it even after their emancipation:

> The rural population, both formerly and at present, recognizes the usefulness of well-organized borzoi hunts and saw in them their salvation from predatory wolves, which began to multiply excessively from the moment that the number of borzoi hunters diminished and borzoi hunting fell into decline.[178]

Fleshing out his explanation of why hunting with hounds had ebbed in the wake of the emancipation, Sabaneev expanded his earlier socioeconomic analysis by specifically linking its decline with aspects of post-emancipation Russia and emerging modernity as they had affected the gentry. He argued that it was not so much that landowners of means—who still existed—couldn't afford to maintain adequate kennels, but rather that many were unable or unwilling to adjust to the social conditions of the post-emancipation Russian countryside and had forsaken it for the capitals or life abroad. In addition, many members of the new generation (the "golden youth" of the 1860s, as he termed them) had begun to view borzoi hunting as antiquated or even "barbaric" as they focused on other pursuits such as the excitement of the stock exchange (*birzhevaya igra*), careers, and various business opportunities.[179] Society, in other words, had changed dramatically in a way more conducive to gun hunting as aspects of emerging modernity came to Russia.[180]

Turning to gun hunting, Sabaneev emphasized that it was inherently more appealing to urban hunters in part because of its more democratic character and the fact that familiarity with horses and dogs was not a prerequisite. In addition, the arrival of percussion caps to Russia in the 1840s and subsequent developments had made firearms more accurate and dependable.[181] Sabaneev then described a gamut of methods of hunting wolves with guns. These ranged from large scale drives, to the Moscow Society's preferred method of employing *pskovichi* who would plan a forest drive in such a way as to force the wolves toward waiting gun hunters, to placing horse meat—or alternatively tying a live young pig or puppy in a remote location—to attract wolves, who could then be shot by hunters lying in wait, to other more specialized techniques. His description of the various techniques was so detailed that he was effectively providing a procedural manual for those who might wish to employ any of them. As in his discussion of borzoi hunting, his description of gun-hunting techniques touched on ways in which the emancipation had affected them, as well as how peasant attitudes played a role in their success or failure. An amusing example of his advice touched on which sorts of peasants served as the best beaters, hired to walk through the woods driving any wolves toward the waiting gun hunters:

> If the population is completely unfamiliar with drive-hunting, then twelve to fifteen year-old-boys make the best beaters because they will follow orders obediently. Adult male peasants smoke and stand around, women gather in groups, but boys will do what they're told.[182]

After exhaustively portraying a variety of gun hunting techniques Sabanaeev finished his book by outlining the approaches utilized by professional hunters and trappers to vanquish wolves. These ranged from running them down on skis over deep snow to utilizing snares, steel traps, and poison hidden in the carcasses of dead livestock. He argued that the government should provide more encouragement and guidance to commercial hunters to incentivize the killing of wolves, noting that a skilled and motivated professional could discover as many as five to ten dens in the springtime, which could lead to a substantial reduction in wolf populations.[183] He also contrasted the widespread custom among the peasants of European Russia of leaving known wolf dens alone out of lack of initiative or fear of retribution with the determined destruction of them by the tribespeople of Central Asia and indigenous peoples of Siberia—the Kirghiz and Siberian Tatars, for example.[184] Having explained the issues that undermined the interest of professional hunters and trappers in focusing on wolves, he recapitulated the problematic motivations of the gentry and peasantry, noting as well the government's complicity in their overpopulation in much of European Russia:

> At the present time young wolves are preserved by peasants because of superstition and egoism and by landowners for their own or other's hunting. The government itself contributes to the increasing numbers of predators because in the less forested provinces—where wolves choose state forestlands as their main place of habitation—de facto the peasantry does not have the right to enter the forest, even without an axe or gun.[185]

The Rise of Hunting Societies, the Professionalization of Wolf • CHAPTER 2 | 57

Figure 8 D. A. Sabaneev, With the Spoils of the Hunt (March 1880). *Nature and Hunting.*

Over the course of the 1880s the Wolf Problem to which Sabaneev had devoted his monograph continued to be debated and became intertwined with a larger discussion of the need to revise Russia's hunting laws in order to bring them into conformity with the new socio-economic conditions of post-emancipation Russia, current zoological thinking, and Western European precedent. A broad-based consensus emerged that game populations had dramatically diminished over the preceding decades due to economic development, deforestation and other changes to Russia's natural landscape, illicit hunting practices, and predation. Turkin, who succeeded Sabaneev in the early 1890s as editor-in-chief of both *Nature and Hunting* and *Hunting News*, provided one of the most significant voices in the debate over reforming Russia's hunting laws. Starting in the late 1880s, as he sat on successive government commissions charged with evaluating and revising the empire's hunting codes, Turkin began to publish an extended series of extremely detailed and well-researched articles on hunting law, which was a subject of perennial interest for the readership of both the monthly journal and the weekly newspaper. Many of these were subsequently incorporated into his three books on the topic, the earliest published in 1889 and the last in 1913.

Turkin's 1889 book—*Hunting Laws: A Critical Study of Russian Hunting Legislation*—first appeared as a series of editorials in *Hunting News* in 1888 and then in revised form as a 220-page supplement to *Nature and Hunting*, as had Sabaneev's monograph on wolves nearly a decade earlier.[186] In it he attempted to systematically analyze Russia's hunting laws and situate them in a comparative European context, emphasizing areas in which Russia lagged behind and should emulate Western Europe (especially Germany and the Scandinavian countries). His analysis was meant to educate readers about the considerations that government committees were taking into account as they worked toward revising these laws, a process which would culminate in the legislation of 1892. Three years later he followed this with a 150-page analysis entitled *The Hunting Law of 3 February 1892*.[187] The work aimed to provide a thorough context for the 1892 law, as well as to explain each of its articles and the reasoning that had led to them. Almost twenty years later, Turkin published a broad overview totaling more than 200 pages of Russian hunting law from its earliest iterations to the present entitled *Hunting and Hunting Legislation during the 300-Year Period of the Romanov Dynasty*.[188] Each of Turkin's works demonstrates extraordinary depth of research, his analytical and comparativistic approaches, and a clear understanding of the challenges that confronted Russians as they attempted to formalize and normalize practices that many Western European countries had implemented much earlier. These included effective policing of closed springtime seasons and of poaching more generally, as well as systematic campaigns to reduce or eradicate predators and especially wolves. Turkin's work touches on some of the same issues that concerned Sabaneev in promoting these approaches in Russia: the interrelationships between peasant, professional, and sport hunters; the role of hunting societies in protecting game species, including gamebirds, as well as reducing predator populations; the inconsistency and inadequacy of Russian hunting laws and their enforcement; and the ways in which Russia lagged behind Western Europe. His focus on the legal aspects of these interrelated questions complements Sabaneev's emphasis on wolf zoology and hunting techniques, with their critiques of Russian "backwardness" providing a point of common emphasis.

For present purposes, I will focus on just two aspects of Turkin's extremely broad-ranging contributions to Russian hunting law: his arguments that Russia lagged behind Western Europe in its hunting legislation and his recommendations for changing Russian laws and policies aimed at predator control and reducing wolf populations in particular. For the sake of coherence and sequentiality I will start with his 1889 book, then bring in certain perspectives from his commentary on the 1992 hunting law, and utilize his 1913 book primarily to fill in any gaps—for example to cite his further research on the results of wolf control efforts in the intervening years.

While noting that Europe differed from Russia in many respects and that hunting law and practice varied substantially among individual European countries as well, Turkin nevertheless argued that Russia lagged behind Western Europe as a whole in its approach to regulating hunting. Moreover, whereas the number of hunters in Western Europe had generally decreased in recent decades, the emancipation had led to a dramatic increase in the number of hunters in Russia.[189] This increase had brought the inconsistencies and inadequacies of the empire's hunting laws into sharp relief:

> Notwithstanding the enormous significance of the hunting economy for our government, we do not have a correctly formulated system of hunting laws. The disorderly state of our hunting economy and the shocking inadequacy of legal statutes and measures of preservation and prevention in Russia stand out with particular force and relief when juxtaposed comparatively with those in Western Europe.[190]

Turkin stressed that Russian hunting statutes imposed an artificial unity across a vast variety of landscapes of differing flora, fauna, climate, and topography, as well as varying cultures.[191] For example, the same prohibition against hunting game birds from the beginning of March to the end of June that had been imposed in 1763, shortly after Catherine II's accession to the throne, still applied across most of the Russian empire.[192] This did not account for the fact that the timing of spring migration varied substantially over the empire's differing geographic regions, as did human population densities and the role of hunting in local economies and indigenous cultures. Ironically, Turkin noted, Russia could benefit from the fact that many species of birds, and of course other animals, spent their entire lives on its territories even during migration. With appropriate policies, he argued, this should enable the empire to exert greater control over its wildlife populations and hunting economy than was possible in the much smaller countries of Western Europe, where migratory birds typically crossed national borders each year, incentivizing the citizens of each country to kill as many as possible when they had the opportunity.[193]

Turkin devoted a dozen pages of the 1889 text to a discussion of the inadequacies of Russian hunting law concerning predatory mammals and birds.

He pointed out that Russia's legal codes specified that "predatory beasts and birds may be exterminated by any possible means."[194] At the same time, he noted, the legal definitions of predators did not accord with current zoological knowledge, as they were incomplete and sometimes inaccurate. According to existing law, for example, mammalian predators were defined as follows: "Predatory mammals include bears, wolves, lynx, foxes, marmots, *and others*" [Turkin's italics].[195] This limited and imprecise categorization, he noted, included a non-predatory rodent—the marmot—but excluded, as could be ascertained elsewhere in the legal code, the predatory polecat.

Turkin discussed the means by which predators were hunted in Sweden, France, and Prussia, noting that the governments of those countries actively encouraged and in some cases required their citizens to participate in predator control or eradication. He focused especially on Sweden as an example of effective anti-predator laws and practices. He emphasized that Swedish hunting clubs were required to hunt predators and that rural Swedes of all classes were obliged to support these efforts and were subject to monetary fines if they did not participate in scheduled hunts led by state-appointed huntsmen.[196] He specified fourteen methods by which Swedes were required to mount organized hunts against predators, as well as to utilize poison, noting that certain methods were allowed only in particular districts.[197] These legal and social norms and carefully-constructed government policies, he concluded, had produced extraordinary results. In certain years Swedish anti-predator campaigns had yielded close to 10,000 bears, wolves, foxes, and lynx killed; this amounted to one large predator for every twelve Swedes. In Russia, by contrast, a typical yearly tally of wolves, lynx, and foxes killed came closer to one predator for every 300 Russians.[198] Turkin summarized the differences between Swedish and Russian practice in the more extended version of the passage from which I excerpted the epigraph for this chapter:

> In Sweden we observe the unceasing development and perfecting of methods for the extermination of predators . . . the active dedication of the government and specialists to the issue of the eradication of predators, a heartfelt acceptance on the part of the people and authorities of the undoubted necessity of the systematic and well-regulated destruction of predatory animals . . .
>
> In both Asiatic and European Russia there is no such desirable unity between the organs of governmental power and the population in the matter of predator eradication, no good-willed recognition of the necessity for cooperative and assiduous work in a great variety of forms

and directions oriented to the destruction of predatory animals, and an absence of serious legal measures lends little hope for a positive resolution in the nearest future. In Russia each of us in disorder and at random, without communication, cooperation, and help defends himself from predators using his own personal and accidental means and methods, which are indicated in our law by the terse term "all possible" (*vsevozmozhnye*).[199]

As evidence of the pernicious effects of such neglect Turkin cited statistics that attested to the toll that Russia's numerous predators and especially wolves took on its rural economy and population, focusing on the years following Lazarevsky's 1876 brochure and drawing on a variety of sources ranging from journalistic accounts to the bulletins of rural *zemstvo* councils. After noting the losses of livestock to wolves incurred by specific rural communities, he compiled reports by year of attacks by wolves in the periodical press in the 1870s and 1880s. In addition to numerous attacks by rabid animals, he asserted that as many as 200 people were attacked and eaten each year by non-rabid wolves in European Russia. He claimed, for example, that 177 people had been killed and eaten by wolves in the representative year of 1883.[200]

Figure 9 Russian family attacked by wolves (1845). *Illustrated London News.* Antiqua Print Gallery/Alamy Stock Photo

Turkin rounded off his discussion by citing the extensive coverage of methods of wolf hunting and eradication that had appeared during the previous fifteen years in the pages of *Nature and Hunting* and listed fourteen techniques currently employed to hunt wolves in Russia (a subset of the gamut of methods described ten years earlier in such detail by Sabaneev).[201] He concluded that this plethora of possibilities indicated that not a lack of expertise but rather a deficiency of government coordination and support lay at the root of Russia's continuing Wolf Problem:

> Only the government can render such help: injunctions concerning the mandatory conduct of legal hunts for predatory beasts, prizes for slain predators, strict control over their eradication, aid to peasants in organizing hunts, etc.—all these are the obligatory means which must be adopted in order to bring about an improvement in our hunting economy and growth in the peasants' standard of living.[202]

Three years after Turkin's 1889 study the law of February 3, 1892 incorporated many of the underlying principles he had advanced.[203] The law establish a variety of closed spring seasons for different bird species that also exempted certain parts of the empire where professional or indigenous hunters predominated, particularly among Russia's northern and eastern provinces. It required that hunters above the age of seventeen purchase an annual hunting license (again exempting portions of the empire). It established a series of fines for varying types of infractions including poaching and trespass, assigning enforcement to local police and other authorities, and offered complete protection to the rare European bison (*zubr*). It affirmed the right of peasant communes to grant hunting permission to individuals within their own communities or to others of their choosing. It prohibited the destruction of non-predatory birds' nests and eggs and various methods of snaring and netting birds, as well as the springtime sale of gamebirds.

The 1892 statute also strongly reaffirmed the Russian legal code's previous bias against predators, codifying the collective cultural animus against these "harmful" species. The list of predators was expanded to include seventeen genera and species of mammals and fifteen of birds. The other predatory mammals on the list, some of which like squirrels (and bears, for that matter) are omnivores that will opportunistically eat eggs and fledglings, included: red and gray foxes, jackals, badgers, Arctic foxes, polecats, weasels, otters, minks, ermines/stoats, martens, wolverines, rats, and wild felines. The list of predatory birds included: eagles, falcons, merlins, hawks, magpies, ravens, crows, jackdaws, jays, nutcrackers, shrikes, eagle-owls, owls, and sparrows (which will kill nestlings and destroy the eggs of other bird species).[204] The statute incorporated the views of Turkin and other experts in that it explicitly addressed the roles of both poison and hunting societies in predator control, indicating that the government should help to direct their efforts:

> Predatory animals and birds, their fledglings and nests, as well as stray cats and dogs, may be exterminated throughout the year in the fields and forests, by every available means except poisoning.*
>
> *Note: Provincial and district authorities are authorized to permit the use of poison as a universal measure for the extermination of predatory animals or to permit individuals or hunting societies to employ it.[205]

In his extraordinarily meticulous book-length explanation of the 1892 law Turkin spent several pages elucidating the considerations that had gone into categorizing mammalian and avian predators, citing perspectives up to the ministerial level. He noted that it was recognized that certain animals, both predators and others, could play either positive or negative roles in different environments and circumstances. For example, polecats served a useful role in parts of southern Russia, where they helped to control gopher populations. In a different vein hares could wreak devastation on agricultural plantings and gardens although they fell under the category of useful, rather than harmful, animals. Other animals such as wolves and bears, however, were considered to have only negative impacts and were therefore to be destroyed as widely as possible.[206] Yet he also noted that, despite this consensus, there was less unanimity about the extent to which the government should fund predator-eradication efforts. In the final outcome a great deal of leeway was left to local authorities in deciding how vigorously to pursue and how generously to fund efforts to reduce predator populations.[207]

From a modern ecological perspective, the 1892 law seems extraordinarily short-sighted in its rigid categorization of animals according to their perceived utility, on the one hand, or the harm or danger they could pose to humans, on the other. It completely ignored the possibility that predators might play a significant role in larger natural systems that extended beyond the narrower sphere of human concerns and interests. As Ekaterina Pravilova points out in her discussion of the law in the larger context of Russian attitudes toward property and natural resources, "the notion of utility in the rural Russia of the late nineteenth century differ drastically from modern representations, and it was more grounded in the tally of human victims and economic losses in man's encounter with the wild."[208] Yet as Turkin himself knew well from his extensive study of foreign hunting legislation, such blanket hostility to predators and especially wolves characterized predator control efforts of that era in other countries as well. In addition to the European wolf reduction efforts highlighted earlier, American predator eradication programs were gaining momentum precisely in these decades and would extend well into the twentieth century, as I will discuss in the conclusion.

The 1892 law remained Imperial Russia's most significant piece of hunting legislation up to the time of the revolution, notwithstanding subsequent attempts to modify and extend it. Despite its emphasis on reducing populations of predators, however, imperial Russia's wolf-eradication efforts never achieved the level of overall coordination or state-level funding that Sabaneev, Turkin and others had envisioned in their aspirational comparisons with Western European countries like Germany and Sweden. Rather, wolves continued to be targeted primarily by members of the larger hunting societies such as the Moscow Society, as I described earlier, and the Imperial Society with its extensive network of provincial branches, as well as in a variety of ad hoc ways at the local level, where hunting associations and clubs often played a role.

In his 1898 history of the Imperial Society, Turkin emphasized the significance of its provincial branches in predator reduction efforts, noting that those in central and southern Russia tended to rely on hound and gun hunting, whereas those in the northern and eastern provinces tended to favor the use of poison. He singled out the predator-control activities of the Novgorod-Tver branch from 1891 to 1897 as exemplary, noting that its members killed an average of eighty wolves and foxes and thirty bears each year. He concluded:

> This branch showed in word and deed that the [Imperial] Society of hunting must not consist of an exclusive corporation closed off from anything exceeding the bounds of the necessities of sport. It clearly understood that hunters join into a society not in order to isolate themselves from life with its burdens, demands, and requirements and to devote themselves to sporting pleasures but rather in order to more successfully come to the aid of those in need.[209]

Despite such efforts Russia's wolf populations remained robust throughout the imperial period. In his 1913 study of hunting legislation throughout the

Romanov era, written on the eve of the empire's demise, Turkin emphasized that Russia's anti-predator efforts still lacked adequate regimentation and that large tracts of privately owned land continued to provide havens for wolves because their owners deliberately protected them for their own hunting or because they did not want to provide access to others to pursue them.[210] From his vantage point in 1913, in other words, the very situation that Sabaneev had highlighted in his 1880 monograph still prevailed. As a result large predators, primarily wolves, continued to pose a threat to the rural inhabitants of Russia: "According to the data of the Central Statistical Committee, from 1870 through 1887 1,445 people were eaten by beasts throughout the 49 provinces of European Russia and 1,200 people from 1888 through 1908."[211]

* * *

Given the centrality of Russia's hunting societies and clubs to its wolf control efforts, such as they were, I will close this chapter with a literary portrayal of a provincial hunting club's wolf hunt that will help to provide texture to the more historical sources explored above. The prose writer and dramatist I. A. Salov (1834-1902) published in a number of Russia's leading journals from the 1850s through the 1890s. His depictions of rural life gained particular respect among critics, who compared him with the more famous I. S. Turgenev. Salov's fifteen-page story entitled "Wolves" depicts a weekend gun hunt in the waning decades of the nineteenth century organized by a hunting club in Saratov, which is located about 500 miles southeast of Moscow.[212] "Wolves," which draws on Salov's experiences as a landowner and hunter who had returned to Saratov after a bureaucratic career in Moscow, is strikingly consistent with both Sabaneev's and Turkin's analyses. As do they, Salov emphasizes the contrast between Russian and Western European approaches to wolf control and the relatively disorganized and ad hoc nature of Russian efforts. He portrays the club's members, its head huntsman, the wolves themselves, and a peasant who intrudes on the hunt in ways that corroborate and complement the non-fictional sources. His story also provides another window on the issues of gentry interactions with the peasantry that we've encountered in this chapter.

Salov's unnamed first-person narrator opens the story by informing his readers that the manager of the hunting club to which he belongs has invited its members to participate in an early September wolf hunt in a forest located in the vicinity of their club's hunting lodge fifteen versts from the city, where a den of wolves has been pinpointed. He and another member of the club hire horses and a carriage for the trip to the lodge, as he fends off good-humored criticism of his aging muzzleloader, which does not compare to his companion's breech-loading Lancaster rifle. He sketches the makeup of their small society, which consists of fewer than forty hunters including "lawyers, notaries, commissaries, civil servants, landowners, railway officials, bankers, Russians, Germans, and apothecaries."[213] Only one of their members utilizes borzois, which he unleashes at any prey that make it through the cordon of gun hunters positioned to intercept animals driven from the forest by the club's pack of scent hounds. In addition to the scent hounds each member's twenty-five-ruble yearly dues provide for the maintenance of the hunting lodge, a few horses, a head huntsman, and a houndsman. He continues:

> You couldn't describe the Saratov hunting society as properly organized. It's sooner a circle of enthusiasts, who aren't dedicated to any goals other than the satisfaction of their own hunting needs. The society doesn't in the least resemble those societies and circles that I've chanced to encounter in various places in Germany. There a hunting society is devoted to more universal goals. It rents forests and fields, has its own watchmen, helps to enforce legal hunting, prosecutes the taking of game out of season—in a word, it provides the larger society with a certain recognizable benefit, uniting the useful with the pleasant. And in truth I've been convinced of the usefulness of those societies on more than one occasion. Wolves don't exist there, they don't fall upon villages in packs, they don't reduce the peasantry to ruin with their depredations, whereas all the game that brings benefit to mankind, rather than harm, is preserved with pedantic exactitude.[214]

Salov's criticism of the Saratov hunting society's self-interested motivations seems ironic, in light of Turkin's earlier rhapsodic description of the Novgorod-Tver branch of the Imperial Society, but corresponds to Turkin's overall lament about the lack of effective predator control in Russia in his 1913 history written fifteen years later. His narrator's praise of German hunting clubs, which he links directly with the absence of wolves in that country, also corroborates Sabaneev's and Turkin's analyses, although his closing reference to their "pedantic exactitude" hints that Russian disorderliness and self-interest has its own attractions (among them the opportunity to hunt wolves).

When they arrive at the lodge, the narrator asks if he may accompany the club's hired huntsman, Adrian, on his midnight expedition to confirm that the wolves are still located in the vicinity of their den (this was a common practice prior to wolf hunts). The two set off on horseback by moonlight to the forest, ten versts away, as distant hills stand out against the night sky and the signal lights for the railway line that reached Saratov in 1870 blink much closer at hand. They tether their horses to trees on the outskirts of the forest and proceed quietly on foot. The narrator mimics Adrian's every move until they find themselves deep in the forest, barely breathing, aware of glowing eyes just a short distance away. Adrian whispers "Wolves!" as he sits up on one knee. Then he adds: "We are seeking them, and they found us. Did you see them?"[215] After a few minutes, the wolves move away. The narrator asks whether they were in danger, to which Adrian responds: "As long as you lie quietly, why would they attack?... Now if you run away from them, that's another story and it's possible they'd go for you... But lie quietly—and they won't touch you for anything."[216]

After this initial encounter the two proceed to the edge of a glade deep within the forest. There the narrator witnesses Adrian's communion with the wolves as he howls with consummate mastery—eliciting and answering a chorus of howls from the pack—so that his imitation and their howling blend indistinguishably into one another:

> This howling, which began with barely audible notes that seemingly arose from underground and from somewhere far away, gradually strengthened and finally transformed into that plaintive, disconsolate wailing, which almost evokes compassion and which stuns you to the depths of your soul. The most despairing notes could be heard as they resounded, everything was here: hunger, and despair, and yearning ... [in the] concert of these predators, who don't have a right to live.[217]

Salov's description of the wolves' howling is strikingly evocative. Just as Tolstoy's Danilo and the other serf huntsmen we encountered in Chapter 1 mediated the gentry's relationship with the natural world, so does Adrian provide Salov's protagonist with an intimate encounter that he could not possibly achieve on his own. Moreover, so skillful is Adrian's imitation that Salov's gentleman hunter is unable to distinguish between the vocalizations of his human guide and those produced by two adult and five young wolves who emerge from the forest and howl only twenty meters from the narrator's hidden vantage point before disappearing again into the night ten minutes later: "Adrian imitated [the old wolf], and those two voices again fell into conversation, again they exchanged questions and answers."[218] Salov's narrator's sense of awe at the encounter resonates with his recognition of the hunger and anguish expressed in the wolves' voices, which Adrian can so expertly mimic and which "almost evokes compassion" for "these predators, who don't have a right to live." Salov's portrayal of this auditory communion, which nearly causes his protagonist to question his culture's attitude of unremitting hostility toward wolves, also closely resembles passages from this period in which looks exchanged between wolves and humans elicited similar uncertainty, as I shall explore in Chapter 4.

The two hunters return to the lodge at three in the morning and set out together with the other members of the hunting club just a few hours later. By eight o'clock all the gun hunters have drawn lots and Salov's narrator is in his place as the scent hounds are loosed in the forest. The hunt itself, however, unfolds anticlimactically in part because of a lack of coordination between the hunters and peasants who are cutting trees in the area. At its outset he hears the sound of an axe and sees a group of peasants down in the ravine where the den is located as the hounds are loosed. As he waits, one of the peasants unexpectedly clambers up to him and settles beside him for the duration. Giving in to circumstances, the narrator engages the intruder in conversation and learns that he is grateful the hunters have come as the wolves have attacked two of his horses, although he vents his true spleen on the merchant from whom he is renting land on unfavorable terms: "The wolves have taken over, and the landlord is gnawing... That one is even worse than the wolves!"[219]

This conversation echoes the very parallels that Sabaneev had asserted between the toll exacted by Russia's wolves from its peasantry and their exploitation by the country's landed gentry prior to the emancipation, although an avaricious merchant now plays the role of adversary alongside the wild predator in the post-emancipation era. For Salov's narrator, the peasant's story of exploitation by both antagonists supplants the excitement of the hunt that he had anticipated, and he focuses instead on learning more about the man's circumstances. By the morning's end four young wolves have been taken—three by gun and one by the borzois—but he hasn't fired a shot. The adult wolves and the remaining pup have escaped, slipping past one of the other hunters in a scene that leads Adrian to lambast him in the same outraged way that Danilo confronted the old Count Rostov in *War and Peace*.

Just as the borzoi wolf hunt had reflected the mores of an earlier era, so the hunting societies of Imperial Russia's waning decades captured the dissonances and tensions of a new age in which the forces of modernity coexisted with age-old tensions between the privileged and the peasantry. Salov's story, which so strikingly parallels the analyses of Russia's Wolf Problem by experts like Sabaneev and Turkin, demonstrates how consistently these aspects of Russian society's coexistence with wolves manifested themselves across a range of sources during these years. The wolves whom Salov's narrator encountered in such proximity, guided in doing so by the hired huntsman Adrian, represented a constant that united the older and newer faces of Russia. In the next chapter, I will explore the ways in which another aspect of modernity—modern medicine—affected the age-old relationship between Russians and wolves in the years leading up to and immediately following Pasteur's invention of a rabies vaccine.

CHAPTER 3

Chekhov's "Hydrophobia," Kuzminskaya's "The Rabid Wolf," and the Fear of Bestial Madness on the Eve of Pasteur's Panacea

> At the present time we look upon a man who has fallen ill with rabies simply as an unfortunate whose fate—with treatment or without—will be the very same, but we don't consider him a walking nidus of infection, the slightest contact with whom can prove fatal for us. In former times, however, the most primitive attitudes toward the contagiousness of rabies prevailed. People were afraid of becoming infected just from spending time in the same room as the afflicted.[220]
> —**A. Kh.**, "About Rabies in People and Animals,"
> *Nature and Hunting* (1880)

Among the life-threatening scourges ranging from cholera to tuberculosis that afflicted Russians during the nineteenth century, rabies was one of the most feared. Usually transmitted through dog bites but most terrifyingly carried by rabid wolves, "canine madness" (*sobach'e beshenstvo*) inevitably caused death among those who exhibited the telltale symptoms of hypersensitivity, despondency, spasms alternating with lethargy, tormenting thirst exacerbated by an inability to swallow liquids, and finally collapse and unconsciousness shortly before death. In the decades leading up to Pasteur's 1885 development of a vaccine, which could prevent rabies in those who had been bitten but not yet manifested the disease, accounts of attacks by rabid animals and especially wolves on people as well as depictions of rabies' horrific symptoms in the face of ineffectual treatments were a recurrent preoccupation in Russian culture, as were reports of its transmission through dog bites elsewhere in Europe.[221] They appeared both in the popular press and in the specialized journals published by Russia's hunting societies, the Russian Society for the Protection of Animals, and the empire's growing medical establishment.[222] They also featured in literary works of the day.

Reports of such encounters reached a crescendo in the story of nineteen Smolensk peasants who were attacked by a rabid wolf on February 17, 1886. Promptly dispatched to Paris for treatment by Pasteur, they garnered considerable attention in both the Russian and French press.[223] Pasteur helped to sponsor their trip for humanitarian reasons and also to further assess the virulence of wolf bites, which he suspected might require a modified regime of inoculations. Ultimately, three of the group died while sixteen returned to Russia. Such "pilgrimages" to prevent the previously untreatable disease soon led to the establishment of rabies stations throughout the Russian empire.

This chapter explores the cultural significance of this nexus of wolves and rabies in late nineteenth-century Russia drawing on diverse sources including journalistic accounts, medical research, and literature. It probes the tensions that characterized relationships between the inhabitants of rural Russia and the medical professionals who were responsible for treating them, yet were unable to do so effectively prior to Pasteur's discovery, as well as the complex role that folk healers played within this dissonant medical landscape. It culminates in a discussion of two literary works—Chekhov's "Hydrophobia" (*Vodoboiazn'*) and Kuzminskaya's "The Rabid Wolf" (*Beshenyi volk*)—which appeared in 1886 in the months after Pasteur's discovery and which diverged dramatically in their treatments of these issues. At the chapter's end I will return to the story of the Russian peasants who journeyed to Paris in search of treatment for their wolf bites, helping to spur the subsequent spread of rabies treatment facilities to Russia itself. Together, these interlinked topics will provide another window into the ways that wolves reverberated within the nineteenth-century Russian imagination.

On March 17, 1886, just over two weeks after Pasteur publicly attested to the effectiveness of his vaccine in treating 350 patients, the writer-physician Anton Chekhov (1860-1904) published "Hydrophobia" in *The Petersburg Gazette*. It was one of the last stories to appear under the pseudonym of A[ntosha] Chekhonte early in this watershed year during which Chekhov accepted his role as a significant writer. Chekhov revised it extensively in the 1890s and the revised version was published in 1905, the year after his death, as "The Wolf" in the journal *Russian Thought*.[224] In each version Chekhov's gentleman protagonist is bitten by a rabid wolf while on a hunting trip in the countryside. He turns to both folk healers and medical doctors in the hope of avoiding the disease and broods morosely about its likely course, encapsulating central themes in Russians' fear of rabies, or *lyssophobia*, in a narrative space situated mostly on the eve of Pasteur's vaccine. He oscillates between optimism and despair as he responds to varying depictions of the disease by a local doctor and others, revealing the power of these medical narratives over his outlook.

In June 1886, just three months after the appearance of "Hydrophobia," Tolstoy's sister-in-law, T. A. Kuzminskaya (1846-1925) published "The Rabid Wolf" in *The European Herald*.[225] Kuzminskaya's story, which Tolstoy read and approved prior to publication, explores the same cultural preoccupations with different emphases. It portrays the struggle between a rural widow and her children with a rabid wolf, the townsfolk's fear of the disease and those who may carry it, and the victims' desperate attempts to avoid developing rabies. These include a horrifying encounter with Russia's emerging medical establishment in the provincial capital, which Kuzminskaya juxtaposes to the spiritual and psychological support lent by a folk healer. Kuzminskaya's emphasis on the female protagonist's perspectives and fate inscribe Russia's larger cultural narratives about wolves and rabies in a dramatically gendered space, while highlighting themes of provincial ignorance and social injustice.

In his influential essay "Framing Disease," Charles Rosenberg provides an interpretive framework that will aid in the challenging task of interrelating these diverse literary and non-literary sources. In Rosenberg's formulation:

> Disease is at once a biological event, a generation-specific repertoire of verbal constructs reflecting medicine's intellectual and institutional history, an occasion of and potential legitimation for public policy, an aspect of social role and individual—intrapsychic—identity, a sanction for cultural values, and a structuring element in doctor and patient interactions.[226]

My treatment will reflect each of these aspects with a particular emphasis on the power of narrative both in literary form and as an expression of power and dominance—or alternatively solace—in the healer–patient relationship to which those in fear for their lives are especially susceptible. In this sense, as will be especially evident in probing Kuzminskaya's portrayal of medicine in the provincial capital, my use of Rosenberg will sometimes tread close to Michel Foucault's emphasis on the ways in which institutionalized medicine has served as a tool of domination and oppression of the marginalized and especially the insane.[227]

I opened the book with a journalistic account of a rabid wolf that attacked a Belorussian village late one night in January 1862, biting fifty-eight people before it was killed. We'll now return to that account and delve more deeply into its aftermath—including both the medical treatments that were administered by the physician in charge, Dr. Grabovsky, and the patients' own efforts to seek help—in order to gain a better sense for the interdependence of medical and folk approaches to dealing with rabies in Russia just over two decades before Pasteur's vaccine.[228] As Grabovsky reported to his colleagues at the Vilnius Medical Society some months after the attack, his treatment involved careful cleaning of the wounds along with application of a poultice of arsenical solution (an approach advocated by his contemporary N. A. Arendt, to whom I will return). While he concluded that it had provided no apparent benefits, he followed the best practices of burgeoning medical science by keeping meticulous records of the disease's progression and his patients' outcomes. Forty-one individuals developed rabies and died while twenty-two survived (he included the five additional victims that the wolf had bitten prior to entering the village in his tally). The first victim showed symptoms twenty days after the attack and the majority over the following month. A few succumbed only several months later. The afflicted exhibited symptoms for an average of three days before dying.[229]

After providing this medically-grounded synopsis of Grabovsky's results, the article then delved into a further portrayal of the human drama inherent in the tragedy along with the desperate patients' attempts to seek unofficial avenues of treatment in folk remedies. Some of the afflicted bowed to the inevitable and bid farewell to their families with stoic resignation, the article recounted. Others, bereft of any confidence in the medical profession represented by Grabovsky and his colleagues, in whose perspectives the anonymous account was firmly anchored, fell prey to charlatans who appeared on the scene offering a range of folk cures:

> After the first fatality a crowd of charlatans appeared with indubitable remedies for the disease and, notwithstanding the watchfulness of the doctors, medical aides and attendants, they snuck in to see the afflicted, who—in despair—strove with all their might to meet with them, which it was difficult to forbid them from doing as they had given up on turning to the medical profession for help. What the healers gave to the ill we don't know, as they concealed it. One of them had brought several types

of bread on which were inscribed talismanic words that would supposedly charm the disease. Another advised obtaining the heart of a killed wolf, searing it on coals, then grinding it into powder and giving it to the victims in vodka ... In light of such a large number of victims from this unfortunate event it becomes grievous to recollect that science has been unable to provide a cure for the bite of a rabid animal up to this day.[230]

The article's closing lament eloquently captures the sense of frustration that accompanied attempts to help those bitten by rabid animals prior to Pasteur's vaccine. A wide spectrum of methods was employed against rabies in nineteenth-century Russia, many of them shared with Europe. These ranged from folk remedies through various medical treatments, none of them effective and few even palliative, which were marked principally by their variety and frequent harshness in the 100 years leading up to Pasteur's vaccine.[231] While neither medical nor folk treatments are deemed effectual in the 1862 narrative, it draws a clear contrast between the secretive charlatans endeavoring to profit from the dire situation through folk remedies rooted in religious or pagan worldviews and medical professionals like Grabovsky, who attempted to analyze it through the emerging methods of scientific inquiry and to share this knowledge with their peers in the medical and scientific communities. In this sense, Grabovsky and his colleagues correspond to Rosenberg's formulation of "physicians as articulators and agents of a broader hegemonic enterprise."[232] Grabovsky's meticulous summary of the percentage of victims who developed the disease (typically much higher in the case of wolf attacks), his enumeration of the varying periods of dormancy that preceded its manifestation, and his overview of symptoms were all characteristic of the medical profession's attempts to develop a rational framework for analyzing rabies and to experiment with possible remedies, which would continue up to and beyond Pasteur's creation of a vaccine.

During the hundred years preceding Pasteur's clinical experiments, which epitomized meticulous application of the scientific method, dozens of articles, books, and pamphlets were published in Russia on the topic of rabies. I will touch briefly on significant examples dating from 1780, 1840, 1859 and 1880 to provide a medical context for Chekhov's and Kuzminskaya's literary treatments. In Rosenberg's terms, these texts will indicate some of the ways in which early medical discourse in Russia "framed" the disease of rabies prior to Pasteur's vaccine, as well as the horror that rabid wolves provoked in the Russian countryside.

The earliest book-length treatment of rabies in Russia that I have discovered dates from 1780. It was written by a prominent physician, Danilo Samoylovich (1746-1805). Samoylovich trained and practiced in Moscow, St. Petersburg, Paris and elsewhere in Western Europe and was known for his work on the bubonic plague. He submitted his fifty-page treatise on rabies along with a companion piece on venomous snake bites from Leiden at the request of the president of the Moscow Medical Chancellery, A. A. Rzhevsky, and it was published in Moscow by the printing house of N. I. Novikov.[233] Quickly republished in a second edition in 1783, Samoylovich's treatise resembled the mid-nineteenth-century examples that would follow in claiming to present a cure for rabies but differed from them in its emphasis on the danger that its victims, bereft of their capacity to reason and resembling the animals that had bitten them, could pose to those around them:

> Of the numerous illnesses that oppress humankind on a daily basis, it would scarcely be possible to seek out one more horrifying and pathetic than to see a man who has been infected by the poison from a mad dog's bite... Having forgotten the human reason endowed him by the Almighty Creator, he embodies all the actions of the dog in the most horrendous manner: he bays and gnaws anything he comes across. He spews a terrible foam out of his mouth and flings himself ferociously upon all who are nearby, including even those who of their own good will try to care for him, or give him food, and tries to bite them with his teeth and greedily to wound them. To say it in a word, being ill in this way a father will not spare his son, nor the son his father, the mother her daughter, the sister her brother.[234]

As a way of emphasizing the danger that a person with rabies could pose, along with the mistaken notion that the disease could sometimes lie hidden for many years (both central elements of folk beliefs about rabies that would prevail into the nineteenth century), Samoylovich cited the story of a newlywed couple. The groom had been bitten by a rabid dog as a child but manifested the disease only on their wedding night: their horrified families found his corpse alongside the mutilated remains of his young wife the next morning.[235]

Samoylovich's portrayal of rabies treads close to the folkloric realm of vampires and werewolves that had transfixed Europe earlier in the eighteenth century and that continued to resonate among European and Slavic rural populations long afterwards.[236] It conflates the demoniacal with the bestial and the sexual, emphasizing above all the savage ferocity of those suffering from the disease. Despite this emphasis on their bloodthirstiness, Samoylovich called for more humane treatment of the afflicted, revealing the stirrings of more modern medical sensibilities. He counseled his readers not to kill those suffering from rabies—for example by smothering them to death with pillows—but to bind them tightly at the first sign of the disease's emergence.[237] In addition, he recommended a series of prophylactic measures. Bite wounds should be treated by either cutting away wounded flesh or cauterizing the area with hot metal (both intensely painful measures). Alternatively, the blood could be allowed to flow freely followed by careful cleaning. For at least ten days a compress of Neapolitan mercurial paste should be applied to the wounds twice daily. In most cases, Samoylovich claimed, this would prevent the disease.[238]

The two mid-nineteenth-century sources to which I shall now turn differed from Samoylovich's publication in deemphasizing the danger that rabies patients posed to those around them (as did most of their medical and scientific contemporaries) but shared his claim to prescribe treatments that could prevent the disease. In 1840 M. P. Maroketti, the physician attached to St. Petersburg's Theatre School, published a 200-page work entitled *A Practical and Theoretical Treatise on Hydrophobia, Including a Preventive Treatment against Rabies*.[239] Unlike Samoylovich, Maroketti emphasized that those suffering from rabies were unlikely to attack relatives or other caretakers; in fact, their torment was often increased by concern for loved ones.[240] Drawing on three decades of experience treating those bitten by rabid animals, Maroketti recommended cauterizing bite wounds while noting that peasants and others of scant means would often simply apply urine and salt to bites.[241] He also claimed that an approach rooted in Tatar folk healing, which he'd employed for more than twenty years, could prevent rabies. His method involved vigilant observation of pustules which he asserted would appear under a patient's tongue prior to full-blown manifestation of the disease: these should be cauterized with hot metal and then treated with an ointment made from Spanish fly. Maroketti claimed that assiduous application of this treatment to approximately eighty bite victims between 1813 and 1838 had prevented all but two of them from developing rabies.

Twenty years later, in 1859, N. A. Arendt published a 34-page pamphlet advocating the method of treatment with arsenic that would be employed and then discounted by Grabovsky and his colleagues in the 1862 wolf attack described above.[242] Arendt, who was from a prominent family of physicians, had undergone medical training in St. Petersburg and then returned to the

Crimean city of Simferopol', where he served as chief medical inspector of the Taurida Province. He described numerous cases of rabies he had encountered in nearly fifty years of practice, emphasizing the prevalence of wolves in the province. He observed that wolf bites were more dangerous than those of any other rabid animal, citing a case in which eleven of twelve people bitten by a rabid wolf had died. He asserted that this stemmed not only from the severity of wolves' bites but also from the greater strength of their "poison."[243] He recounted several cases in which those bitten by wolves had perished after seeking treatment from local healers (*znakhari*), who had prescribed the internal and external application of herbs including greenweed (*drok*) and an ointment made from the blister beetle (*maika*). Arendt claimed that from the mid-1830s to the late 1850s he had successfully treated numerous patients bitten by rabid dogs and wolves through a two-month regime combining meticulous care of the wounds and the application of poultices based on an arsenical solution for which he provided the recipe.[244] Arendt also noted that he had never heard of a case in which one human had infected another with rabies.[245]

These representative texts demonstrate some of the principal contours of the ways in which early medical practitioners in Russia framed the disease of rabies and suggested ways to prevent it. Prior to Pasteur's vaccine, which quickly gained widespread acceptance among the international medical community, competing theories about how rabies was transmitted and illusory cures maintained standing in part because outcomes varied so greatly after a rabid animal's attack. Many of those bitten by rabid dogs and even wolves simply didn't develop symptoms, which could give credibility to the claims of both folk healers and physicians that their methods had prevented the disease. Prior to widespread acceptance of the scientific method exemplified by Pasteur's experiments—and given the fact that rabies' viral mechanism remained a mystery for decades even after his development of a vaccine—it was not surprising that folk and professional medicine both overlapped and diverged in their approaches to the mysterious illness, advancing competing narratives within the contested terrain of rabies treatments.[246] The line between folk healing and institutionalized medicine in general was more porous in the eighteenth and for much of the nineteenth century in both Western Europe and Russia, whereas very soon after Pasteur announced his vaccine physicians advocating other methods of treatment were likely to be labeled as frauds.[247]

Even prior to Pasteur's discovery, however, a broad consensus had taken shape among the scientific and medical communities in Russia and abroad that no effective treatment for rabies had been developed. An erudite twenty-page article that appeared in *Nature and Hunting* in 1880 delved into the state of medical understanding of rabies in Russia just two years before Pasteur began his rabies experiments. Its author, identified only as A. Kh., but almost certainly a physician, analyzed perceptions of rabies in Russian culture, addressing both folk beliefs and recent medical research. He portrayed the intertwined physiological and psychological progression of the illness, providing further context for Chekhov's and Kuzminskaya's literary portrayals of 1886. He also underscored the significance of Russia's large wolf populations in comparison with the more domesticated environs of Western Europe:

> There they are dealing almost exclusively with dogs. In our case, aside from an enormous number of dogs, we also have countless multitudes of wolves and other wild animals. And that the latter play a role in this we can ascertain simply from the meager newspaper reports that reach us.[248]

A. Kh. then outlined current scientific understanding that rabies could be transmitted only through contact with the saliva of a rabid animal, typically through bite wounds, rather than arising spontaneously (a common misconception at the time). Agreeing that wolf bites posed the greatest risk, he added that the incubation period following a bite would typically last from eighteen to sixty days but could end sooner or last as long as six months or even a year. Among simple folk, however, the belief prevailed that the sickness might emerge after nine days, nine weeks, or nine months and that until the ninth year had passed one should not feel safe.[249]

At the end of this hidden phase the individual would begin to experience the disease's first, sometimes ambiguous, symptoms. A. Kh. then described the entire course of the disease in clinical yet evocative language:

> At times the person who has been bitten experiences odd sensations within himself, and describes them variously: sometimes as a sound, or a sharp pain, as a burning, or a feeling of being nipped. He begins to complain of a headache. An abnormal spiritual discomfort grows within him: he becomes despondent, although unable to explain to himself the reasons for this despondency... These precursors take hold for a day, three at the most, and then the period of irritation which truly marks the disease of rabies commences... Breathlessness, spasms, particularly of the throat, a heightened degree of receptivity within all the sensory organs and terrible sensitivity: these are the characteristic types of attacks.... Add to this

hallucinations with meaningless raving, paroxysms... [The sick] try to defend themselves from imagined offenses and insults, rush upon those who are nearby in a rage, hit and insult doctors and those who are caring for them. The sick are not in the least malicious during these fits, as others have said earlier; sometimes they demand that those nearby keep at a distance in order to avoid being bitten. But now the fit has passed—they recognize their relatives and acquaintances, begin to ask for their forgiveness, ask not to be left alone, ask them to pray... All the symptoms of the previous time manifest themselves feebly to be replaced by an absolute absence of strength: the harbinger of approaching death.[250]

This powerful passage eloquently narrates the disease's progression from an initial state of unease to its full-blown manifestations and fatal outcome. The ambiguity of early symptoms, along with the enormous variability in the period of latency and uncertainty about whether the condition would develop at all, helps to explain the disquiet that a person bitten by a rabid animal must have felt in the years prior to Pasteur's vaccine. In Rosenberg's terms, A. Kh. interpreted the "biological event" of rabies for his readers both culturally and medically, elucidating the interlocked physiological, psychological, and even spiritual torment so characteristic of the illness. Along with most of his nineteenth-century peers, he emphasized that people with rabies posed little danger even in their most frenzied states, belying the beliefs of earlier generations. The remainder of the article outlined the variety of treatments typically employed against the disease, noting that none of them had proven effective. Only one thing was certain: if a person exhibited symptoms, death would invariably follow.

The writer-physician A. P. Chekhov's "Hydrophobia (A True Story)" appeared in *The Petersburg Gazette* on March 17, 1886. Both in its original version and as revised and published posthumously, the story demonstrates the literary fecundity of many of the issues discussed above, which Chekhov utilized to explore themes of human hubris and vulnerability in the context of an encounter with a dreaded predator and disease. The story approaches rabies as a liminal mystery that lies on the border of the animal and human, the biological and social, the known present and unknown future in ways that Rosenberg's theoretical approaches will help to elucidate. It pays particular attention to the power of narrative in representing disease and to the role of narrative in the complex relationship between doctor and patient.

The academic edition of Chekhov presents the 1905 version, "The Wolf," as authoritative, including the original publication via variations as part of the scholarly apparatus. While the two versions share many common elements, Chekhov also introduced substantial changes: he altered his description of the protagonist Nilov's encounter with a rabid wolf, the portrayal of rabies in the story, and its temporal flow and scope. Significantly, he excised a reference to Pasteur from the earlier version during his later revisions as part of shortening and changing the story's conclusion.[251] My discussion will straddle the two versions, noting points of commonality and difference and focusing on some of the most salient. Ultimately, this approach will illuminate the ways in which Chekhov's story reflects the medical and social context of rural Russia through its juxtaposition of folk, gentry layperson, and medical discourse about rabies and wolves, as well as the ways in which a provincial doctor enunciates and interprets the latest medical developments to a country gentleman whose fate hangs in the balance after his fearsome struggle with the rabid wolf. It will demonstrate that Chekhov's physician faithfully articulates the medical knowledge of his day but also tries to give hope to his patient; this leads him to alternately frighten and comfort Nilov through his pronouncements about rabies in ways that evolve significantly between the earlier and later versions.

At the outset of the story as originally published in *The Petersburg Gazette*, Nilov, a landowner and chairman of the local governing council, is returning from a day of bird-hunting on foot. He is accompanied by two companions, the examining magistrate Kupryanov and the local doctor Pegasov. They stop for the evening at a millhouse inhabited by an old peasant, Maksim. Maksim asks them to lend him a gun, as a rabid wolf has been in the area for two days and has bitten two dogs. Maksim has seen the wolf and expresses his fear of the animal in terms that reflect folk beliefs in the supernatural: "he looked at me like Satan and gnashed his teeth . . . I was frightened to death."[252] Maksim's identification of the wolf with Satan (*satana*) and emphasis on its fearsome teeth tread close to the folkloric territory of vampires and werewolves mentioned earlier. Dismissing Maksim's fears as superstitious, the gentry hunters reassure themselves that they have weapons. Nilov boasts that he would simply use the butt of his gun on the wolf rather than wasting birdshot, recollecting a time when he put down a rabid dog with just a stick. As they sit over tea, vodka, and cognac Pegasov embarks on a narrative that combines his medical background with a story-teller's desire to grip his audience:

> "There's no disease more torturous and terrible, gentlemen, than hydrophobia . . . A healthy man is walking along, carefree, thinking

about nothing, and suddenly out of nowhere a rabid dog snaps at him! In a moment he's possessed by the terrible thought that he's perished irrevocably, as if he's tumbled into an abyss or fallen under a train. After that you can imagine the agonizing, oppressive anticipation of the illness, which doesn't leave the bitten one for a single minute. After the anticipation follows the illness itself." ...

Pegasov began to describe the symptoms of rabies. I've noticed that doctors, especially young ones when they are carried away, love to describe illnesses, without stinting on the richest details, with complete fervor. They get carried away, relish it, their eyes light up as if the conversation were not about illnesses, but rather about a beloved woman or the beauties of nature. Such an oddity can be explained not so much by a dulling of their sensibilities through habituation, as by a youthful and passionate attachment to their subject. Moreover, those unshakeable laws by which diseases proceed, and indeed the very depictions of suffering, are not without their own sort of poetry ...

"Worst of all is that the disease is incurable," Pegasov finished. "Once someone has come down with it, you can write them off. There's no cure. The medical field hasn't even hinted at the possibility of treatment. Of a hundred who get sick, exactly a hundred will die."[253]

Doctor Pegasov's narration exerts a powerful effect on his audience. In contrast to Maksim's folkish fear of the rabid wolf as a demonic creature and Nilov's smug gentry confidence that he could dominate and vanquish the wild animal as he did a domestic dog, without even firing his weapon, Pegasov focuses on the awesome finality of the illness in medical terms. He compares the anticipation of rabies and succumbing to the "unshakeable laws by which it proceeds" with the terror of falling into an abyss, pulled downward by the inexorable force of gravity, or beneath a train, that symbol of modernity that had only recently come to Russia's provinces. A first-person narrator, hidden elsewhere in the story, steps forth at this one juncture to emphasize the doctor's poetic passion for his subject. In Rosenberg's terms, Pegasov frames the biological disease of rabies as a powerful symbol of the contingency of human fate, employing symbolic language that reflects the era: modernity may have brought trains to Russia, but it has not yet provided a cure for the scourge of rabies, against which medicine remains helpless, and which the country's most feared predator can carry in its saliva.

Maksim alone among the group counters this pessimistic portrayal with the claim that local folk healers can treat rabies effectively:

> "But in the village they treat it, sir!" said Maksim. "Miron will cure anyone you like."
> "Nonsense," sighed Nilov. "As for Miron, all that's just tales. Last year in our village a dog bit Stepka and none of these Mirons were any help ... No matter what rubbish they had him drink, he still went mad."[254]

In his later revision Chekhov eliminated the character of Pegasov, dividing portions of his dialogue between Kupryanov (now Nilov's sole hunting companion) and a district doctor named Ovchinnikov, who appears only in the story's final pages.[255] Kupryanov—a magistrate rather than a physician rhapsodizing about a horrifying and compelling challenge to his profession—tells the others about rabies. Correspondingly, the later version omits the entire first-person passage that anchors the initial description of rabies in a youthful doctor's passion for his field of expertise, which likely reflected Chekhov's own fresh memories of medical school (he had graduated in 1884). Chekhov also replaced the conjoined references to the laws of gravity and the forces of modernity with a religious term not present in the earlier version. Rather than Pegasov's statement that a man bitten by a rabid dog feels "as if he's tumbled into an abyss or fallen under a train," Kupryanov says "as if there's no salvation" (*net spaseniia*). Moreover, Maksim's reference to the rabid wolf as "Satan" is replaced by the similar but less loaded "evil spirit" (*nechistaya sila*).[256]

In both versions the hunters are left brooding about human mortality as they continue to sit over their drinks:

> The horrific tales about rabies had their effect. The company gradually ceased speaking and continued to drink silently. Each involuntarily pondered the fateful contingency of life and human happiness upon chance and trifles that would seem insignificant and not worth, as the saying goes, a hollowed-out egg. A sense of boredom and mournfulness came over them.[257]

Their ruminations link the danger of encountering a rabid animal with the overall randomness of existence. Given the futility of treatment, the disease symbolizes death itself, along with the anxiety and suffering that precede it.

Later, Nilov blithely goes out for a walk—inexplicably (but perhaps drunkenly) leaving his gun behind—and encounters the rabid wolf. Hoarsely groaning, it zigzags towards him in the moonlight:

> On the weir, awash in moonlight, there wasn't a bit of shadow; in the middle the spout of a broken bottle shown like a star . . . But suddenly it seemed to Nilov that on the other side, above a bed of willow, something resembling a shadow moved in a dark circle. He squinted his eyes. The shadow disappeared but soon showed again and moved in zigzags along the weir.
> "The wolf!" Nilov remembered.[258]

Their paths converge as if they must and moments later the two are locked in a struggle to the death as the strongly built landowner attempts to hold the salivating and desperate creature at arm's length while the wolf struggles both to attack him and to escape. In both redactions Nilov's encounter with the wolf occupies nearly a full page of text. Like an 1872 passage in which Tolstoy fictionalized his own mortal struggle with a bear he had been hunting, Chekhov's passage focuses on the raw physicality of the encounter, so different from Nilov's earlier experience with a frenzied dog.[259] As a rabid predator that emerges from the darkness and shadows, the wolf represents animal nature in its most fearsome form. Yet the third-person narration, anchored in Nilov's perceptual point of view and closely attuned to his anguished thoughts, also implies a degree of empathy and even kinship with the crazed creature. The 1886 version is particularly explicit in juxtaposing Nilov's consciousness with the animal "other" while emphasizing that their shared vulnerability unites them:

> Both, Nilov and the wolf, whose heads were at the same height, looked into each other's eyes . . . The wolf snapped its teeth, let out rasping noises, and spit saliva . . . Its hind feet, seeking support, scrabbled at Nilov's knees . . . Its eyes, in which the moon shone, didn't express evil. They cried and resembled those of a human. What did the sick wolf feel?
>> All of Nilov's strength, both of his muscles and nerves, went into his hands. He didn't think, felt little and just strove to hold on . . . All the details, even the most minor, of the terrible portrayals the doctor had sketched appeared before him, though momentarily. In the face of the awful present there was no time to think about the past or the future.[260]

The contrast between human and animal, so proximate in the present and potentially so interlinked in the future though the transmission of the deadly disease in the wolf's saliva, evokes a visceral truth about mortality and suffering that crosses the barrier between species. As they struggle, Nilov's physical sensations and mental distress coalesce into a feeling of existential terror that he will experience the torment he sees reflected alongside the moon in the wolf's eyes, so vividly described by Pegasov and so unlike the diabolical menace feared by Maksim. While the unsettling howling of wolves under the moonlight was a recurrent motif in Russian culture, Chekhov utilizes Nilov's awareness of the moon's reflection in the eyes of the wolf, so close to his own, to assert a powerful but enigmatic link between the two beings.

In the later version Chekhov excised the entire second paragraph with its explication of Nilov's thoughts. He also substituted the anatomically more appropriate "paws" for "feet" and truncated the last sentence of the first paragraph, leaving only: "The moon shone in its eyes, but nothing resembling evil could be seen there; they cried and resembled those of a human."[261] As in his elision of the first-person frame surrounding Pegasov's impassioned description of the disease, this excision leads to a sparer and more neutral narrative; yet it maintains Nilov's recognition of the wolf's suffering and its commonality with his own terror. It may also reflect the fact that Kupryanov, rather than Pegasov, had described the effects of rabies in the later version.

Nilov's desperate cries attract his companions and soon the wolf lies dead on the ground, apparently choked to death. Nilov's own shoulder, however, has been deeply torn by the wolf's jaws. In both versions Nilov spends a sleepless night and then begins a frantic series of consultations and treatments in the hope of avoiding rabies. Despite his earlier derisive comments about folk medicine, he follows Maksim's advice to swallow a spoonful of a loathsome liquid while facing East and then visits the local healer he earlier disparaged, Miron. Chekhov does not enter into the details of Miron's methods, but the alert reader is aware of the irony in Nilov's turning to him. Chekhov's greatest attention, however, is devoted to Nilov's interactions with the local doctor, who first provides him with tablets made from the herb belladonna and then focuses on providing psychological reassurance. He goes so far as to read his patient a chapter from a medical text about hydrophobia, although Chekhov notes that he omits the scarier passages. Pegasov (in the later version Ovchinnikov) also comforts Nilov by citing statistics and probabilities, as well as evidence drawn from the encounter:

"You have a much better chance of not falling ill, than falling ill. In the first place, only thirty of a hundred who are bitten get sick. Then it's also very important that the wolf bit you through clothing, which means that the poison remained on your clothes. Even if poison did enter the wound, then it would have had to flow out with the blood, since you bled profusely."[262]

The doctor's reassurances are consistent with the medical knowledge of his day. In his 1887 archival analysis of 693 patients admitted after rabid animal bites to Moscow's Ekaterinskaya Hospital, for example, the physician and researcher D. P. Kishensky concluded that about thirty percent of those bitten by rabid wolves developed rabies.[263] This proportion exactly corresponds to the doctor's estimate in Chekhov's story, although it is lower than the figures advanced by most of Chekhov's contemporaries, including Pasteur. It may also be worth mentioning in this context that the district doctor Ovchinnikov in Chekhov's 1888 story "An Unpleasantness," mentioned above, "was known among his colleagues for minor works on medical statistics."[264]

In the 1886 version Nilov takes some comfort from the doctor's counsel but nevertheless grows thin and gray and falls prey to mysticism over the following months. His fear of the dread disease alone is enough to age him prematurely. The story culminates in an ironic twist typical of Chekhov's early work that refers to Pasteur's newly developed vaccine:

> Nilov did not fall ill with rabies. A month passed, then a second and a third, and little by little he began to liven up. After a year even the encounter with the wolf would have been forgotten, if not for the reminder of his prematurely gray hair and wrinkles.
>
> People have poor memories. Not long ago a rabid dog bit the old man Maksim. Pegasov, without giving it much thought, set off for Nilov. "We're sending Maksim to Pasteur," he told him. We're soliciting donations. Wouldn't you like to contribute?"
>
> "Oh, with pleasure!" Nilov left, and a few moments later handed the doctor *ten* rubles.[265]

The concluding passage of the 1886 version of the story refers directly to Pasteur's treatment as a potential remedy for the disease that Pegasov had described as incurable a year earlier in narrative time when Nilov was bitten by the wolf. This narrative interlude allows Chekhov to explore Nilov's fear of developing rabies without any possible recourse. The irony of Nilov's stinginess is heightened because the reader remembers his desperate cries for Maksim to come to his aid during the attack, as well as the fact that he had offered the doctor an enormous sum to cure him (100,000 in the 1886 version, reduced to 50,000 in the later revision) when confronted with his own possible illness.[266]

The later version of the story ends only four days after the attack and there is no direct mention of Pasteur's vaccine. Instead, comforted by the doctor's assurances that circumstance and probability indicate he won't develop the disease, Nilov leaves Ovchinnikov's office in high spirits. He regales his peasant coachman with the tale of his struggle with the wolf, embellishing the account with colorful detail as he laughs merrily and remarks that it will give him something to remember in his old age. While his optimism may be misplaced it impresses Dr. Ovchinnikov, who admires him as he leaves, thinking: "What a hero! . . . What a fine one!" The informed reader knows, however, that Nilov may well develop rabies, as the incubation period extends past the narrative's limited time frame. Perhaps the doctor's admiration for Nilov in Chekhov's revised ending implies that optimism in the face of our common mortality should be valued rather than viewed as a form of hubris. Or perhaps the revision highlights the same misplaced arrogance that led Nilov to boast that he could vanquish a rabid wolf with the butt of his gun prior to the actual encounter. In either case, Chekhov's decision to eschew any mention of Pasteur in the later version contrasts Nilov's optimism with the dire and irremediable fate that will await him if he should sicken. The obvious irony of the 1886 version, which juxtaposes Nilov's premature aging from existential dread of the disease with his stinginess as Pegasov arranges for Maksim to benefit from the newly invented panacea, transmutes into a subtler form of irony in which Nilov denies the possibility that he will succumb to rabies, though in fact he may.

Kuzminskaya's "A Rabid Wolf," which appeared just three months after "Hydrophobia," provides a strong counterpoint to Chekhov's depiction. Whereas Chekhov's portrayal focuses on Nilov's perspectives as a member of the gentry and foregrounds and privileges professional medical discourse about rabies over folk beliefs, Kuzminskaya's story embodies distrust of the state's emerging medical institutions while emphasizing the kindness of a capable folk healer who helps an uneducated widow to navigate difficult circumstances and lends her psychological support. Yet it also underscores the ways in which rural Russians ostracized and mistreated those bitten by rabid animals, as evinced in some of the medical sources above. In addition, fully a third of her narrative is devoted to a description of the wolf's attacks on other townspeople and the widow's family, attesting to the lurid fascination such accounts held for readers of the day and providing a contrast with Chekhov's portrayal of Nilov's solitary encounter with a rabid wolf.

Subtitled a "true story" (*istinnoe proisshestvie*), akin to Chekhov's subtitle a "true event" (*byl'*), the seventeen-page narrative was based on an actual occurrence familiar to both Tolstoy and Kuzminskaya. In a mid-October 1885 letter addressed to his sister-in-law, Tolstoy reacted enthusiastically to the draft:

> I've just read and approved your manuscript. During the first free evening I'll reread it with a pen in hand and with stern judgment. But it seems to me that it will be necessary to correct very little: the story is very interesting and simply written. It's a pity that you didn't write about how the police harassed her—the district police officer demanded and they ordered that she bury the wolf and kill the dog. It seems that's how she told the story.[267]

Tolstoy's response emphasizes the real-life harshness of the district authorities as conveyed in Kuzminskaya's literary portrayal, and indeed Kuzminskaya's account utilizes the violent onslaught of a rabid wolf to explore the ways in which provincial ignorance and bureaucratic injustice increase the suffering of her widowed protagonist and her six children in the wake of its attack on their inn near a forest and railroad station in the Tver' province.

While the major narrative focus of the story is on the widow Anna Ivanovna and her children, the wolf first falls on two men on a nearby road: Alexander Gerasimovich, a mining clerk who sometimes hunts with Anna's oldest son, and a railway official, Nikolai Fyodorovich. The two men are able to temporarily beat the wolf off and make their way to a nearby cottage where another German worker lives with his cook Matryona and her husband, Ivan. No sooner have they arrived than the wolf, which has followed them, attacks Ivan, knocking him off his feet and biting him and his wife before fleeing.

A bit later that same winter evening Anna Ivanovna's eldest son, seventeen-year-old Vasily, sees the wolf chasing the beloved puppy given to him by Alexander Gerasimovich around their isolated yard. He runs impulsively to its defense only to be knocked off his feet by the wolf:

> "Mommy, a wolf is biting me! Mommy, help me!" Vasily cried. But the wolf was biting his face and hands, which Vasily was using to protect himself. As he bit his hands, Vasily drew them back. Then the

wolf would bite him in the face, chomp once, bite hard, and let go. In this way the wolf was biting his hands, and cheeks, his nose, and his chin. His mother ran up and saw: "Heavens! Good Lord! Vasya!" she shouted and with bare hands, without an axe or stick she rushed at the wolf, as if it were a lamb. She fell onto the wolf and began to struggle with it.

"Mikhail, Misha!" she called. "An axe! Quickly an axe!" Misha was already running out of the cottage and without heeding his mother's words saw only that both his mother and brother were struggling with the wolf. And he, like his mother, ran with bare hands straight to the spot. He jumped onto the wolf from above and grabbed hold of its throat with both hands... As soon as Anna Ivanovna saw that the wolf had opened its jaws she pulled her arm out from under the wolf and thrust it into the open jaws.[268]

The struggle continues for a full two pages until the eldest daughter, Natalya, arrives with an axe:

Natalya brandished it and deftly struck the wolf on the head, but at that very moment she screamed: "Mommy! I've scalded my face!"... Splashes of hot blood had drenched her face; she hadn't expected this and in the cold air the warm blood seemed as hot as boiling water. The wolf's head had been severed and the corpse that had caused so much harm now lay sprawled on the snow covered entirely in blood.[269]

As in Chekhov's depiction of Nilov's encounter with a rabid wolf, the lengthy passage emphasizes the brute physicality of the family's struggle with the crazed and powerful creature, as well as its sheer bloodiness. Anna Ivanovna and her children sacrifice themselves heedlessly for each other, eventually prevailing over the wolf. Weaponless, Anna Ivanovna thrusts her arm into the wolf's mouth (exactly as the villager had done to protect his pregnant wife in the 1862 newspaper account cited earlier); this enables her daughter to utilize the axe. The scene's climactic moment—in which the wolf's hot blood spills over Natalya as she hews it—juxtaposes her virginal innocence and courage with the beast's bloodlust in a folkloric flourish.

Having devoted the first third of her narrative to describing the wolf's assaults on the family and their neighbors, Kuzminskaya embarks on a detailed description of the onslaught's aftermath, focusing on the ways in which provincial ignorance and fear impact the victims as they variously succumb to or avoid developing rabies. An overarching theme—which echoes her famous brother-in-law's distrust of institutionalized medicine—is that local practitioners and folk healers provide more comfort to the victims then do their professional counterparts in the city, whom she represents as callous and misguided. The forces of modernity—from the railway that transports the victims to the provincial capital at the insistence of the local factory manager, to the medical professionals who attempt to imprison them in a madhouse there—exert excruciating pressures on Anna Ivanovna and her neighbors, while offering no cure. A folk healer who lives on the outskirts of the city, bridging the rural and urban worlds, provides their only source of solace.

Anna and her family are first treated by a local female medic (*doktorsha*), who cleans their wounds and gives them vodka. Like the district doctor in Chekhov's story, but with less expertise, she tries to reassure Anna by saying that the wolf is unlikely to be rabid—"just a bit, for sure, crazy in some way."[270] Unfortunately, her words have the opposite effect on Anna, who had never before seen a wolf or considered the possibility of rabies:

> "But do rabid wolves really exist?" Anna Ivanovna asked with trepidation. "They exist, but really very rarely." It hadn't entered Anna Ivanovna's head earlier that the wolf might be rabid and the female doctor's words upset her. "Yes," she thought. "It must have been rabid, or why would she have been trying to calm me?"[271]

The manager of the nearby factory, accompanied by the local elder, then arrives by carriage and insists that all those bitten by the wolf take the train immediately to the hospital in the provincial capital. Anna pleads that her children, as young as three, will be abandoned. In response, the manager contributes a meager ruble toward her ticket and threatens to use force if necessary. When they arrive at the hospital after midnight, however, the night porter promptly dispatches them instead to the city's madhouse in an eerily Foucauldian vein:

> "Bitten by a wolf, then it must have been rabid," the porter answered. "We don't admit people with rabies."
> "Then where should they go?" asked the clerk. "To the madhouse," he answered, yawning widely."[272]

Anna and her neighbor Matryona spend a sleepless night at the madhouse amid a chorus of laments that penetrate the walls of their locked room. In the morning a doctor examines them and pronounces that they must remain in the asylum under observation for six weeks, while ruminating that they will surely develop rabies given the severity and location of their wounds. They implore him and the other staff to allow them to return home for the sake of their children but are met with "the same indifference, haste and inattentiveness."[273] Anna realizes that her only recourse lies in subterfuge. She conceals her severely wounded hand from the asylum's senior doctor, who permits her to leave after examining the superficial abrasions on her other hand. She and two others are released, while her son Vasily, Matryona, and the mining clerk, Alexander Gerasimovich, are compelled to remain for observation.

Liberated from the madhouse, Anna visits a sixty-year-old folk healer named Aleksey Semyonovich, who lives in the city outskirts with his widowed sister and whom "all knew as a good healer" (*zagovorshchik*).[274] In contrast to the callousness of the asylum's doctors, he treats her gently and sympathetically as she explains her circumstances:

> "Well, what's to be done, Anna Ivanovna, don't be afraid, we'll assuage your grieve and rescue your son. I'll bring him to you myself." The old man looked over her wounds and went behind a wooden partition. He brought a glass of water and some bread. He cut it into pieces, lit the lamp, and began to whisper prayers to himself with a serious face. Breathing deeply, he blessed the water and bread. After ten minutes he gave this bread to Anna Ivanovna to eat, and some of the water to drink, and washed her wounds with the remaining water. A half hour later Anna Ivanova bid him farewell and set off homeward by train.[275]

Aleksey Semyonovich's methods draw on folk custom and beliefs. The bread that he blesses, and his prayers on her behalf along with his kind demeanor, help to reassure Anna in ways that are analogous to the comfort that Nilov finds in consulting with the district doctor, who embodies such a different worldview. Kuzminskaya's story embeds reassurance in folk and religious custom, whereas Chekhov's narrative employs statistics and physiologically-based observations about the effects of clothing and wound severity on transmission of the disease. In both cases, however, a healer's psychological encouragement takes the place of effective treatment in the era prior to Pasteur's vaccine. In fact, Aleksey Semyonovich's pledge to rescue Anna's son from the madhouse implies that institutionalized medicine as practiced there will have only negative effects on him.[276] In Rosenberg's terms Kuzminskaya's story privileges a traditional healer-patient relationship, based on shared communal and religious traditions, over the inhumane practices of modernity.

Aleksey Semyonovich is able to prevail on the administration of the madhouse to release the others for his treatment and observation. Back in their homes, they continue to seek remedies over the next month from local healers. At this point Kuzminskaya details the final days of those who succumb to rabies in ways that echo the medical accounts detailed earlier while emphasizing the ignorance and resulting inhumanity of the rural population in its treatment of the victims. The numbers three and nine, which carried particular significance in folk beliefs concerning rabies as highlighted in the 1880 article cited above, and the first of which coincides with the typical period of overt symptoms preceding death, occur repeatedly:

> The first to fall sick, after three weeks, was Matryona. She began to feel melancholy, to cry, not to sleep at night and then the attacks of madness truly began. She began to fear water and to snarl in a voice that was not hers. They locked her up at such moments and kept her daughter at a distance from her. And then at a certain point when one of her fits began, they locked her in a cold storage room—there was nowhere else—they tied her up and left her that way. After several hours it grew quiet; they unlocked the door, looked in, and she's lying dead on the floor, already cold and stiff. [Her husband] Ivan grieved, and became frightened: would he end in the same way? But Ivan was fated to get better.[277]

Nine days later the clerk, Alexander Ivanovich, senses that he may succumb. Reflecting an internalization of the same historical attitudes explored earlier, he asks his young wife to fetch others to bind him, saying: "Move away: I'll bite!"[278] Over the course of the next three days, he is alternatively released and tied up again until he develops unambiguous symptoms and dies while en route to the city by train. A similar fate befalls another neighbor, who leaves behind a wife and three young children.

Anna and her son Vasily are more fortunate. They make it through the initial six weeks of greatest danger without developing symptoms and accept the healer's reassurances that they will not succumb to the disease. The wolf's attack, however, has irrevocably altered their lives for the worse. At the end of the story three years have passed. Their once thriving inn has been forsaken by the townspeople and others. "Where are you going?" the simple folk would say to each other. "Go if you want to die: that woman innkeeper is surely about to go mad. And no one went to her."²⁷⁹ Anna Ivanovna has written to the provincial authorities to request a reward for killing the wolf but has heard nothing. The story ends with a final forlorn sentence: "And she only asked for 75 silver rubles."²⁸⁰

Individually and collectively, the fates of Kuzminskaya's characters underscore the ways in which superstition and fear coalesce to isolate and stigmatize those bitten by the rabid wolf. From the factory administrator and village elder to those who forsake their inn, the rural population sees them as tainted and liable to develop rabies long after the incubation period has passed. The wolf itself embodies utter ruination: its attack is a catastrophe from which their family can never recover. Equally exposed to the onslaught of this creature that emerges from the nearby forest and the unforgiving forces of modernity embodied in the city's madhouse, Anna Ivanovna and the other victims exemplify the vulnerability of late imperial Russia's rural population in Kuzminskaya's literary adaptation of a real-life story.

Both Chekhov's and Kuzminskaya's narratives demonstrate the fecundity of rabid wolves and rabies as literary topics situated at the intersection of traditional folk medicine and beliefs and the burgeoning field of professional medical discourse in Russia. The two authors contrast these discursive spheres in very different ways that demonstrate their divergent allegiances amid the changing ecological, social, and ideological landscapes of the post-emancipation era. Each first exploits the drama inherent in a direct encounter with a feared predator, then turns to the existential dread that often overcame those bitten by rabid animals prior to Pasteur's vaccine as well as the types of medical treatments they sought or to which they were subjected. Chekhov's portrayal focuses on the psychological travails of his protagonist and highlights the importance of a local doctor's counsel, while Kuzminskaya pursues a pointed social critique. Viewed against the medical and cultural context laid out earlier, as well as through the interpretive lenses provided by Rosenberg, the two stories elucidate the deep imprint the figure of the rabid wolf left on Russian culture in the age before Pasteur. Whereas Kuzminskaya's narrative contains no hint at the panacea that Pasteur would develop, the original ending of Chekhov's story—which appeared precisely a month after the attack by a rabid wolf on the Smolensk peasants and shortly after they had arrived in Paris—refers briefly to Pegasov's plan to send the peasant Maksim to Pasteur for treatment.

I will conclude this chapter with a brief overview of the story of the Smolensk peasants who were treated by Pasteur after being bitten by a rabid wolf, which helped to pave the way for the subsequent establishment of rabies stations in Russia. The wolf's attack took place on February 17, 1886 in the town of Bely, located in the western part of European Russia in the Smolensk province.[281] This happened to be the very day on which Pasteur read a paper at the French Academy of Sciences in Paris detailing the results of his new vaccine in treating 350 patients over the previous seven months. Among them only one—a girl whose treatment had started too late—had died. The Russian medical community had been following the progression of Pasteur's rabies experiments closely since they began in the early 1880s, providing periodic updates in journals such as *The Doctor*, and efforts quickly began to arrange for the nineteen peasants bitten by the rabid wolf to travel to Paris for treatment.[282] Their journey was delayed as the necessary funds were raised and passport arrangements were made, a process which rose to the level of Tsar Alexander III himself through the mediation of his close advisor, the chief procurator of the Holy Synod of the Russian orthodox church and state councilor Konstantin Pobedonostsev.[283]

The nineteen peasants arrived in Paris on March 1, nearly two weeks after the attack, and commenced the series of inoculations with the attenuated rabies virus that Pasteur had developed through his experiments on rabbits and dogs. Their presence in the city created a sensation, as described in an account by a French doctor published in the *Moscow Gazette* on March 23.[284] He noted that the Russian consul was one of the first to visit the peasants and had promised them his "fatherly" support as the tsar's representative. The citizens of Paris, both poor and wealthy, were also contributing food, money and clothing to the Russians. The author of the article had recently visited them: "I found them remarkably calm and courageously awaiting the outcome of their sorry fate."[285] As for Pasteur, he examined each of the patient's daily as they received their inoculations (in a manner that strongly contrasts with the medical professionals represented by Kuzminskaya):

> I myself was a witness to the particular sympathy that your compatriots receive from Pasteur and, I will confess, it touched me deeply to see so much simple warmth and tenderness in this renowned deity of science in his interactions with these simple folk, torn from their far-off land and lost amidst the city of Paris.[286]

Ultimately, three of the peasants died, while the remaining sixteen returned to Russia after thirty-seven days. Prior to his treatment of them, Pasteur had adamantly opposed the opening of rabies clinics in other countries, believing that relinquishing control over the application of his vaccine might imperil his chances of demonstrating its efficacy and reliability to the medical community. Partly as result of his experience with the Russians, however (and convinced that three had died because of the delay in beginning the series of inoculations), he acquiesced to facilitating the establishment of rabies treatment centers in Russia and elsewhere. Reflecting on his experience with the Smolensk patients, whom he singled out of the thousands he had treated in an 1889 article, Pasteur noted that the bites of rabid wolves were inherently more dangerous than those of dogs:

> This wolf, running through the fields for two days and two nights, fell upon the peasants with such viciousness that he disfigured some and wounded others terribly. This group of Russians caused even greater worry because—while certain statistics indicate that one of six bitten by a dog will die—this percent significantly increases in the case of being bitten by a rabid wolf. The poison is one and the same but a dog, having bitten, runs further whereas a wolf, throwing itself in a fury upon its victim, increases the quantity of the poison that is introduced. Often of twenty people bitten by a wolf all twenty will die.[287]

Figure 10 Russian peasants in Paris for treatment by Pasteur (1886). Reproduced with permission of the Pasteur Museum, Paris.

Pasteur continued by noting that the three Russians who succumbed had suffered grievous head wounds but might nevertheless have been saved had not two weeks elapsed prior to the beginning of their treatment, as all data indicated that minimizing the time that elapsed between the bite of a rabid animal and the beginning of inoculations was paramount. He added that an autopsy of the deceased had revealed that one of the wolf's teeth had broken off inside a victim's skull, indicating that the poison had entered directly into his brain at the time of the attack.[288] Pasteur concluded by noting that inoculation facilities had subsequently opened in several foreign countries including six in Italy and seven in Russia, as well as in other European countries and as far away as Mexico and Rio de Janeiro.[289] Thus, the story of the Smolensk patients figured centrally in Pasteur's own explanation of the importance of establishing an international network of rabies inoculation facilities.

The very first of these foreign rabies station opened in Odessa in June of 1886. This was followed in the same month by the opening of a station in Moscow, and in quick succession that same year with the establishment of facilities in St. Petersburg, Warsaw, and Samara.[290] Together, these stations allowed for the treatment of patients from throughout the Russian empire and beyond. By December 1886, 815 patients had been treated among the five stations, of whom only twenty-two succumbed to rabies. By imperial Russia's demise in 1917 there were 35 stations in the empire, and well over 100,000 patients throughout its territories had received Pasteur's inoculations against rabies.[291]

Despite the widespread acceptance of Pasteur's vaccine and the rapid establishment of rabies stations across the Russian empire, rabies remained a significant concern in tsarist Russia and into the post-revolutionary era. Numerous pamphlets, many of them intended for the uneducated, were published during these decades explaining the disease and providing instructions for how to treat those bitten by rabid animals.[292] Some of these continued to highlight wolf attacks as especially likely to lead to rabies without a prompt regime of vaccination. Yet Pasteur's vaccine, while not always a panacea given the severity of wolf bites, ushered in an era in which the crazed creatures so feared by Russians lost some of their ability to inspire the terror they had possessed in an earlier age. Fear of rabid wolves as incarnations of evil—the superstitious horror that inflected Maksim's description of the rabid wolf as a demonic spirit in Chekhov's narrative and that provided the context for Kuzminskaya's juxtaposition of an innocent widow and her children with a crazed wolf who embodied utter ruination—would gradually give way to a more modern understanding that wolves were both victims and carriers of a virus that caused suffering across species. In the next chapter I will further explore the ways in which Russians' attitude toward wolves evolved at the turn of the century, allowing for the subsequent development of a sense of compassion for these predators among some Russians that Chekhov foreshadowed in his depiction of Nilov's fleeting sense of empathy for the rabid wolf.

CHAPTER 4

Fissures in the Flock: Wolf Hounding, the Humane Society, and the Literary Redemption of a Feared Predator

The she-wolf breathed heavily. Her torn and wounded flanks, on which the red meat could be seen, quivered like a pulse. Her tongue, sticking out from her bloody muzzle, lay on the ground and red froth flowed from it. But, strangely, her eyes did not hold an expression of fierce rage; large and wide, they sooner expressed grief and bewilderment. She looked at the people with a sort of surprise, as if asking why there were people nearby, and not understanding what had happened.[293]

—S. Pospelov, "A Cruel Pastime: Hounding" (1905)

Previous chapters have explored the ways in which Russians demonized wolves and strove to reduce their numbers through borzoi and gun hunts, as well as utilitarian methods including poisoning. Both the anti-wolf activities of Russia's hunting societies and its legal codes as embodied in the hunting law of February 3, 1892 reflected a widespread social consensus that wolves were a scourge to be combated by any means possible. In the closing decades of the nineteenth century, however, and coincident with Pasteur's discovery of a rabies vaccine and the subsequent establishment of rabies clinics in various parts of the Russian empire, isolated voices began to question Russian culture's demonization of wolves and to criticize at least some of the methods used to control them. Some of this questioning occurred among hunters themselves as they contested methods and motivations in their ongoing debates in Russia's hunting journals over the most appropriate and effective means of reducing wolf populations, as I explored in Chapter 2. But the most forceful interrogation of the war on wolves and even hunting itself took place elsewhere both in the popular press and especially in the journal of the Russian Society for the Protection of Animals (*Rossiiskoe Obshchestvo pokrovitel'stva zhivotnym*), which had cultivated an uneasy relationship with hunters since its establishment in 1865.[294] While the RSPA's major emphasis was on protecting domestic animals such as carriage horses and livestock from mistreatment, it also took an interest in the empire's wild game populations. Both hunters and RSPA members, for example, expressed concern over the snaring and netting of live songbirds for the marketplace and the destruction of birds' nests by peasant boys during the spring nesting season.[295] These shared concerns by people who were mostly members of the gentry and professional classes were interlinked with anxiety about the changing nature of the countryside in the wake of emancipation, as apprehension about the future of Russia's wilderness and wildlife coalesced with a recognition that animal welfare and the plight of Russia's rural poor were bound together in complex ways. Despite some points of overlap, however, members of the RSPA were adamant in their opposition to certain aspects of Russia's hunting culture.

The RSPA particularly opposed the practice of "wolf hounding," which many of its members considered wantonly cruel. A series of articles that appeared in its monthly journal focused both on the suffering of the wolves and other animals pursued by competing packs of borzoi hounds in front of spectators and on the pernicious influence of these spectacles on the morals of those in attendance, particularly the young and uneducated. As this indicates, critics were often as troubled by the perceived spiritual and psychological effects of cruelty on its perpetrators or those who witnessed it as they were with the suffering of the animals themselves. Alarm over the morality of the common people amid the changes that followed emancipation also revealed itself in criticism of some of the methods used by rural peasants to kill wolves, ranging from indiscriminate use of poison to various "barbaric" methods that were described in RSPA publications and elsewhere.

Literary and memoiristic works also promoted the glimmerings of a more compassionate sensibility toward wolves. These ranged from Chekhov's 1895 story about a nurturing mother wolf, "Whitebrow," to a series of stories and articles that appeared in the RSPA's monthly journal and elsewhere, some of which were narrated from the perspectives of wolves, to literary works about wolves by other significant writers of the day. Such accounts—some like "Whitebrow" intended for children but others for a broader audience of adults that extended well beyond hunters—attempted to represent wolves' perspectives through various forms of narration and with greater or lesser degrees of anthropomorphism.

Some of the most compelling accounts were by reformed hunters. Tolstoy himself had renounced hunting in about 1880 (as he chronicled in his *First Step*), and his close collaborator, confidante and editor V. G. Chertkov anchored his own renunciation of hunting in an epiphany spurred by a wolf hunt. I will begin this chapter by detailing Chertkov's account, which led to a spirited reaction among both hunters and non-hunters. In tracing this reaction, I will briefly sketch the history of the RSPA from its founding in 1865 to the turn of the century, emphasizing especially its interactions with Russia's hunting community. After describing one especially notorious method employed by peasants to kill wolves, I will delve into the passionate conflict over wolf hounding competitions, presenting the perspectives of both proponents and critics. In the second half of the chapter I will discuss several literary works that presented wolves' perspectives. At the chapter's end, I will examine some of the challenges involved in these attempts, which will help to set the stage for the book's conclusion.

On November 13, 1890, Chertkov's article entitled "A Wicked Pastime" appeared in the St. Petersburg newspaper *Novoe vremia*. Chertkov, whose controversial relationship with Tolstoy has most recently and thoroughly been described in a monograph by Alexandra Popoff, had rapidly gained Tolstoy's affections and support as a spiritual mentor after they met in 1883, when he was twenty-nine and Tolstoy was in his mid-fifties.[296] A former member of the Imperial Guards from one of Russia's most aristocratic families, Chertkov became an ardent advocate for Tolstoy's Christ-centered anti-institutional approach to faith, which he promoted as the editor of the Intermediary publishing house. Intermediary espoused the aim of producing morally uplifting literature at affordable prices for a mass readership and published a number of Tolstoy's works starting in the mid-1880s. Chertkov has become infamous in Tolstoy studies for the ways in which he insinuated himself into Tolstoy's most intimate confidence, usurping the intertwined roles of confidante and editor as Tolstoy's relationship with his wife became steadily more problematic in the wake of his crisis of faith in the late 1870s and early 1880s. I will not delve into that thicket of issues here but will instead focus on a previously neglected aspect of their relationship: Chertkov's authorship and Tolstoy's editorship and support of an anti-hunting article that hinged on a wolf hunt. In early October 1890 Chertkov sent Tolstoy a draft version of the article, noting that he had been inspired to write it by ongoing polemics in the popular press over hunting:

> I'd very much like to get your opinion about what I've written; if it's worth it, wouldn't you point out what should be added to or omitted and in general if you think it's worth including this article in one or another newspaper? Could it be at all useful?[297]

Tolstoy thoroughly edited the draft, shortening and contouring it, as he informed Chertkov in a letter of mid-October 1890: "I just finished correcting your article about hunting. It's very good, because it's useful. I corrected it and omitted a lot . . . I wrote a few words of introduction to give the article a wider reception."[298] In a further letter of October 18, accompanying the revised version as he sent it back to Chertkov, Tolstoy noted that he had carried out further editing, attesting to his avid interest in the topic:

> I'm sending you, dear friend, the article about hunting. At first it seemed to me that I had corrected it well, but then I corrected it a second time and saw that I had corrected it badly. Some things are better, but some things are worse—so correct it yourself without worrying about me. But that it's a good article I won't doubt.[299]

Chertkov's response when he received the edited article indicated that Tolstoy had indeed devoted significant effort to revising the piece. Taken together, these passages from their correspondence also demonstrate that it's impossible to fully identify what portions of the final article represent Chertkov's own language as opposed to Tolstoy's editorial changes:

> I don't even know how to thank you for improving my article. In addition to what it's gained as a whole, and therefore in persuasiveness, your corrections are of great value to me personally as an indication of what is good and weak in the expression of my thoughts. I was touched to the core upon seeing how much labor you put into improving the article and how you almost began to rewrite it yourself.[300]

Partly because of the accompanying preface by Tolstoy, "A Wicked Pastime" quickly gained notice. Tolstoy's preface—written about a decade after he'd renounced hunting — spanned six short paragraphs. In it he told of an old hunter urging a younger hunter to give up hunting, quoting the older man as saying: "It's wrong to kill animals for amusement, not out of necessity." Tolstoy seconded this in his absolute way, saying: "One can neither object to this nor fail to agree with it. It's so simple, clear, and indisputable." Tolstoy added that hunting's days were numbered, as people were waking to this moral truth.[301]

In the following piece of about 3,000 words Chertkov described how his decades of passionate hunting had given way to a sense of profound uncertainty that killing animals—for sport or utility—is justified. After holding his doubts at bay for some time, an encounter with a wolf solidified them. He portrayed the act of bestiality that led to his renunciation of blood sport: a wolf drive that culminated in his beating a wounded wolf to death with a stick as the wolf stared up at him. He recollected the event:

> But once, standing at the edge of a forest during a drive, I got a wolf with a shot and ran up to him to finish him off with a thick stick kept for that purpose. I beat him on the bridge of the nose—the most tender spot on a wolf's body—while the wolf stared at me straight in the eyes with savage rage and gave a deep sigh with each blow. Soon his paws began to twitch convulsively, his limbs stretched out, a light trembling ran through them, and they grew stiff.[302]

Chertkov's description presents a striking parallel with earlier accounts in which vanquished wolves mutely return the gaze of their tormentors, as in Tolstoy's earlier depiction of the bound wolf in *War and Peace*. In this case, however, the encounter caused the hunter to experience an epiphany. On the night of the hunt Chertkov lay awake and contemplated his tormenting of the wolf in an eerily Tolstoyan vein:

> That night as I lay in bed I remembered the day's impressions and in my imagination kept returning to that moment when the wolf showed up at the forest's edge and began to look around. I recalled how the wolf, not having noticed me and hearing the cry of the beaters behind him, was about to run from the forest onto the open steppe, and how at that moment I shot

him and how I began to beat him.... Recollecting this [beating], I realized that I was taking real pleasure in the suffering of the creature as it expired. I felt ashamed of myself. At that moment I realized suddenly, not with my mind but with my heart, that this murder of the wolf was in and of itself an immoral act, that even worse than the act was the enjoyment I took in it, and that worst of all was that lack of principal with which I justified it.[303]

Chertkov described how this epiphany led him to renounce hunting and become a staunch opponent of it, focusing especially on the fact that compassion (*sostradanie*) is one of humankind's most precious qualities and that hunting requires its deliberate suppression. He used visceral language to reinforce this point, emphasizing that hunting desensitizes its participants to the suffering they inflict and undermines the natural human tendency toward empathy:

> To rip open a belly with a dagger, to smash a skull on a stump, to tear into pieces, etc.—all these acts are most common and even necessary while hunting. But surely each person is naturally inclined to have compassion for animals and it's painful to see their suffering. Why then do these very same people, as soon as they are hunting, not only not feel compassion but feel no compunction about deceiving, persecuting, chasing, hounding, tormenting and torturing animals in every way?[304]

Chertkov addressed the utilitarian argument that killing predators is justified as part of his larger rejection of the violence that characterizes hunting on spiritual and even philosophical grounds. In this sense, he was aligning himself with Tolstoy's own emerging philosophy of non-violence and vegetarianism, which he also advocated in the article:

> I realized that if I, while killing the wolf, comforted myself with the notion that I was saving his victims from death then, being in the position of the wolf himself, I could say in just the same way that by remaining alive and eating a hare, for example, I would be saving those insects that the hare would have swallowed together with his food, and so on without end.[305]

The hunting press reacted strongly and defensively to Chertkov's article. One of several pieces that it provoked was printed less than two weeks later on November 26, 1890 in *Hunting News*. It represented editorial opinion, as it was unsigned and appeared in a section of the paper often devoted to editorializing and was most likely written by either Sabaneev or Turkin. After quoting Tolstoy's preface in full and summarizing the sentiment of the longer article it accompanied, it noted ruefully that the famous author's words would surely be making the rounds of Russia and meeting approval among many non-hunters. It then raised objections to the substance of Chertkov's anti-hunting argument, stressing that it would be hard to find a country in Europe in which the population was less protected from the ravages of predators: "Here, every year, with the utmost regularity, up to 200 people are devoured by wild beasts, and hundreds die after being bitten by rabid wolves, who also infect domestic animals with their poison."[306] It argued that the yearly toll predators inflicted on Russia's agriculture amounted to tens of millions of rubles, calling it a "national calamity."[307] After stressing the importance of hunting in combating predators, the author added that Russians should have the right to hunt the wild game that inhabited their lands, as it was part of the country's natural wealth. For all these reasons, the article concluded, hunting and especially hunting wolves in Russia was an important activity that should be supported, rather than simply an idle source of amusement:

> Consequently hunting, properly pursued, serves as a means of self-defense and the defense of others against predators, as a means of maintaining equilibrium among the fauna in the midst of which humankind is destined to live eternally, as a means of supplying the nation with food and—whether hunting takes place for the satisfaction of the sporting passion, for the sake of pleasure, or because of obligation or need—it is equally legal and ethical.[308]

In addition to such reactions in the hunting press, "A Wicked Pastime" attracted the attention of non-hunters, as the editorial in *Hunting News* had anticipated. In January 1891 an article by M. F. Aderkas, entitled "A Wicked Pastime" after the original piece, appeared in the journal of the Russian Society for the Protection of Animals.[309] Aderkas began by noting that one of the various "rebuttals" to Chertkov's piece published in the hunting press had asserted that fewer vicious and brutal people are to be found among hunters than the general population. He questioned this assertion in part by reminding his readers of the historically troubled relationship between hunters and the RSPA. He noted that nearly twenty years earlier—in 1873— hunters had been invited to form a subgroup within the Society in order to cooperate in combating illegal hunting. This relationship, he recollected, had foundered in part because hunters continued to indulge in the practice of baiting and hounding animals, including wolves, which the RSPA considered inhumane and corrosive for the morals of spectators who paid to view these spectacles:

> In 1873 several members of a society of enthusiasts of hunting expressed the desire to join together with the Russian Society for the Protection of Animals in order to struggle together against violations of hunting law and the inflicting of needless pain upon animals. This joining together at first seemed desirable to the RSPA because, not being a hunting society, it didn't have the capability of actively pursuing those violations of hunting law that take place in the hunting sphere. With this goal the charter of a society of hunting enthusiasts as a branch of the Russian Society for the Protection of Animals was constituted and then approved by the Ministry of Internal Affairs. But good words don't always lead to good actions: the hunters continually violated the charter of the RSPA, then seceded from the Society, and finally introduced animal baiting events among their activities. Not possessing any justifying goal, these baiting events represented a vile amusement and means of profit for their organizers, who collected a fairly significant sum from the crowd in support of the baiting. This deprived those who attended of both their wages and the last spark of those feelings, which are inherent to humankind, of compassion for God's creation.[310]

Aderkas's focus on animal baiting (this included competitions in which rival packs of borzois hounded captive wolves, foxes or hares) united several strands in the history of Russia's RSPA, as well as its interactions with the country's hunters, and its particular concern with improving the moral sensibilities of Russia's population. Since its establishment in 1865 the Society had pursued a broad agenda that centered on the protection of carriage and sleigh horses, farm animals, and pets but also encompassed the treatment of stray dogs, slaughterhouse practices, illegal or inhumane hunting practices, and the general goal of inculcating a humane sensibility toward animals among the population, especially the uneducated and youth.[311] In an 1875 overview of the first ten years of the society's existence, V. Iversen included a section on its activities related to wildlife and hunting that highlighted its efforts to reduce the use of leg-hold traps, to forbid illicit snaring and overly tight or crowded confinement of songbirds in cages and their sale on the streets, and to prohibit public display of the feeding of live prey to boa constrictors, which "cannot fail to have an influence on the mass of spectators detrimental to the cause of the protection of animals."[312] Along similar lines, he highlighted the society's success in helping to outlaw performances by trained bears, which often had teeth and claws removed and were subjected to cruel treatment by their trainers. He emphasized that the Society opposed this practice both from concern for the welfare of the bears and because of its tendency to inure spectators to the bears' suffering.[313]

A conviction that underlay much of the RSPA's work was that Russia's lower classes, and especially the rural peasantry, exhibited particular cruelty to animals. Prior to delving further into the controversy over wolf hounding competitions, which were generally organized by hunters from the upper classes, I'll provide one example of a method of controlling wolves in the countryside that encapsulates this concern. In 1893 an article by G. Voronov appeared in the RSPA journal entitled "A Savage and Dangerous Method of Destroying Wolves."[314] Voronov noted that this method had been employed by the peasants of the Novgorod province from ancient times to the present. Hearing the howls of wolf pups from mid-June to mid-July, they would locate the den. The most intrepid among the villagers would then approach when the adults were away, pull out the pups, bind them with ropes, and take them to the nearest river. There they would nail them by their paws to round rafts, which they set afloat down the river. The yelping of the crazed pups was intended to cause the older wolves to pursue them along the river banks. The rafts, being round, would typically float a good distance, clearing the countryside around the den for some time if the method worked.

The first half of Voronov's 1893 description presented a matter-of-fact accounting of this technique. In the second half of his article he turned to a critique of both the practicality and the morality of this approach:

> There's no argument that the wolf is a serious enemy and predator of domestic animals; but surely the important thing is that while tearing, let's say, a sheep into pieces, the wolf isn't acting consciously (*soznatel'no*), not with a preconceived goal of inflicting harm upon humans, but instinctively, for the satisfaction of his hunger. The human, then, knowing this full well, instead of destroying the entire wolf family quickly with a salvo of shots, becomes a tormentor on a level with the unreasoning beast, subjecting him to a terrible death from hunger and the loss of blood.[315]

Voronov's moral outrage at the fate of the wolf pups floating downriver, impaled in a perverse sort of crucifixion for the sins of their elders, reflected the Christian underpinnings of the RSPA. At the same time, his emphasis that the wolves were acting from instinct, rather than intentional malice or evil, showed the growing influence of more modern zoological understandings of animal behavior. In addition to the inhumanity of this technique, he argued that it subjected communities downriver to the danger posed by the enraged adult wolves. Reflecting a common misconception at the time, he conjectured that the distraught adults could pose even more of a threat if they spontaneously developed rabies because of the combination of distress over the fate of their pups and the hot rays of the sun: "and what a rabid wolf means all who live in the country know very well," he concluded.[316]

This approach to wolf control by the peasantry as described by Voronov was a relatively easy target for condemnation by members of the RSPA and gentry hunters as well for both moral and practical reasons, and likewise because it was a manifestation of what they perceived as lower-class ignorance. As a literate contributor to the journal Voronov was not, presumably, vulnerable to the depredations of adult wolves in the same way as were rural peasants. In fact, the peasants may have pursued this strategy in part because—as we have seen—gentry hunters and hunting club members generally waited until the fall borzoi season or winter gun season to target wolves, who meanwhile could carry out their depredations on the peasants' livestock.

Wolf hounding, in contrast, presented a more complex dilemma in the sense that it was typically sponsored by hunting societies dominated by members of the upper classes, as maintaining kennels of trained hunting borzois was a pursuit primarily for the wealthy. RSPA members objected both to the suffering of the animals involved and out of conviction that the public display of animals fighting one another to the death inculcated immorality and cruelty in spectators, who often included women and children. Hunters, on the other hand, argued that coursing competitions served both a sporting and a utilitarian purpose by allowing for the evaluation of the competing packs of borzois. Generally, those opposed to the practice tended to use the term *travlia* (baiting or hounding) to describe it, while those who supported it favored the term *sadka*.[317] As practiced in both capital cities and elsewhere into the 1890s, it involved unleashing borzoi hounds on prey animals—most commonly rabbits and wolves—in order to judge their abilities to run down and incapacitate the quarry. The borzois generally tore the rabbits to pieces, while wolves were sometimes just pinned down and mauled, although they often died in the encounters as well.

An eyewitness account by V. S. Tolstoy, which was published in *Nature and Hunting* in January 1880, provides an introduction to this cultural institution from the viewpoint of a hunter and connoisseur of the practice. Tolstoy began by noting that the competition, sponsored by the Moscow Imperial Hunting Society, was held on January 4 at the Moscow racetrack in the presence of hunters without an audience of the general public. It commenced when a lone wolf, previously captured, was released from a box specially designed so as to expose him on all four sides at once.

> [The first borzoi hound] Ubei caught up with the wolf that had just been released and got hold of his upper leg, but without toppling him. Letting go of him, he nipped him a second and then a third time, then loped alongside him without getting a hold, breaking his stride when the wolf snarled at him. As the wolf was getting away, the pack of V. A. Sheremetyev was released on him. This pack consisted of three males . . . The white one got far ahead of the other two, caught up with the wolf, grabbed him, and the two toppled over together. The other two caught up, grabbed the wolf by the scruff of his neck, and the wolf was pinned. To say that the superb take-down exhibited by these dogs was a rare thing to witness would be weaker praise than they merited. It would scarcely be possible to hold on to a beast—without its getting away and in one place—more ferociously and tenaciously.[318]

As Tolstoy's description implies and in keeping with the borzoi hunts we explored in Chapter 1, the primary purpose of these competitions was to test the speed, ferocity and tenacity of the borzoi hounds, while also training them to confront wolves during actual hunts. The results of such competitions appeared alongside those of horse races and dog shows in Russia's hunting journals and sometimes in the popular press. Descriptions of the *sadki* in publications intended for hunters tended to either tabulate the results in a matter-of-fact manner or, in the case of eye-witness accounts by enthusiasts, to lavish praise on the competing borzois for their respective merits. As in V. S. Tolstoy's account, one detects no hint of compassion for the wolves or compunction at subjecting the captive animals to the borzois' rending jaws.

Non-hunters, however—and particularly members of the RSPA—focused on the institution as a quintessential example of cruelty to animals, and wolves naturally figured prominently as a symbolic element in their rhetoric. Chekhov himself wrote a four-page description of such a competition entitled "At the Wolf

Hounding" in 1882 for the literary journal "Moscow."[319] The account shares certain features with the humorous sketches of contemporary life that Chekhov wrote to pay his way through medical school under the pseudonym of Antosha Chekhonte prior to winning the Pushkin Prize in 1886 and embracing his role as a "serious" writer. Despite its early provenance and informal tone, however, the article emphatically condemned the institution of hounding through the use of biting sarcasm. Chekhov described the event, which—like the one described by V. S. Tolstoy two years earlier—took place at Moscow's horse-racing track in early January, as an "anachronism" in a nineteenth-century European capital, thereby implicitly critiquing Russia's backwardness. Noting that he was not a hunter, Chekhov "apologized" in advance for his lack of mastery of hunting terms and knowledge, saying that he would report on the event "in the same way that the public judges: that is to say superficially and by first impression."[320] He described the large crowd of paying spectators who had gathered at the track, which included women armed with binoculars and school-age children full of impatience for the hounding to begin, then added sardonically: "There are several carts. In the carts there are wooden crates. In the crates the heroes of the day—wolves—are relishing life. They, in all likelihood, are not burning from impatience . . . [321] Chekhov noted that the crowd's nervous excitement built as a wooden box was placed in the center of the arena and they discussed whether the Mazharov or Sheremetyev borzois would be unleashed first:

> They knock on a crate with a hammer . . . The impatience reaches a climax . . . They step away from the crate. One man grasps a rope, the walls of the enclosure fall away, and before the public's eye appears a gray wolf, the most esteemed of Russia's animals. The wolf looks around, stands, and runs . . . The Sheremetyev dogs rush off after him, and behind the Sheremetyev dogs runs a Mazharov dog out of order, and after the Mazharov dog a borzoi handler with a dagger . . .
>
> The wolf didn't have the time to run even two *sazhens* [twenty feet], and already it was dead . . . Both the dogs and the borzoi handler had distinguished themselves . . . and "Bravooo!" cried the public.[322]

After recounting how the same quick death came to a second wolf, Chekhov unleashed his true disdain for this Russian cultural institution in his description of the demise of the third wolf:

> The third crate is exposed. The wolf sits and doesn't stir from the spot. They crack a whip in front of his muzzle. Finally, he stands up as if

exhausted, broken-down, barely dragging his hind legs along... He looks around... There's no salvation! But he wants to live so! He wants to live as strongly as those who sit in the gallery listening to the gnashing of his teeth and looking at the blood. He tries to run, but that's not in the cards! The Svechin dogs snatch hold of his fur, the borzoi handler plunges a dagger right into his heart and—vae victis! [woe to the vanquished]— the wolf falls, carrying with himself to the grave a poor opinion of man ... No joking, humanity has disgraced itself before wolves in contriving this quasi-hunt!... It's another matter to hunt in the steppe, in the forest, where humans' bloodthirstiness can be excused by the possibility of an equal struggle, where the wolf can defend itself, run.[323]

Chekhov's biting emphasis on the wolf's perspective on events, which casts humanity in such a poor light, is highly emotive. He focuses on the wolf's initial immobility and weakness, presumably the result of extended confinement, and its despair at the hopelessness of its situation, in highly empathetic terms. His narration invests emotional force in the wolf's perspective, melding together his own disgust at the hounding with the wolf's desperation and intermingling their points of view. He clearly identifies and sides with the wolf, disparaging the "quasi-hunt" as completely unfair in comparison with a true hunting situation in which the wolf would at least have a chance to defend itself or escape. He saves particular venom for the bloodthirsty crowd of spectators, underscoring its displeasure that the weakened wolf has cheated it of further entertainment and directly linking the hounding with sadism:

> The public is displeased that the wolf was stabbed so quickly... It was necessary first to chase the wolf around the arena for two hours, to have the dogs bite him with their teeth, to trample him with hoofs, and only then to stab him... As if he hadn't already been hounded, caught, and sat for no good reason in his prison for several weeks.[324]

At the close of the article, Chekhov asserted that the baiting served no purpose, as the wolves were too tightly confined before and lacked space on their release to truly test the dogs. While the ticket sales might pay for the expense of mounting the event, he concluded, they could not compensate for the spiritual injury inflicted on the children in the audience, for the "disturbances... wrought by the hounding in the young soul of the schoolchild."[325] In this regard his article squarely echoed RSPA concerns about the effects of such events on spectators.

Figure 11 A Bound Wolf. Lithograph after painting by Aleksei Danilovich Kivshenko (c. 1891). Image copyright Lebrecht Music & Arts

V. S. Tolstoy's and Chekhov's accounts represent antipodes in their approaches to describing these events. As a hunter and connoisseur of wolf hounding events, Tolstoy devotes no thought at all to the wolves' suffering and views the hounded wolves in a purely instrumental sense. Chekhov's diametrically opposed representation instead intertwines the narrator's feelings with his projection of the wolves' despair and the ethical toll that witnessing the hounding exacts on the audience. Members of the RSPA, reflecting sensibilities similar to those that underlie Chekhov's 1882 feuilleton and recognizing the importance of wolf hounding as a symbol of cultural identity (a symbol of Russian backwardness, in Chekhov's account), continued to express their avid opposition to the institution over the following years. Recurrent articles in the Society's monthly journal, speeches given at Society meetings, and fictional works all castigated it throughout the 1890s and into the first decade of the twentieth century.

A five-page article that appeared in February 1890 in the Society's *Bulletin* is representative of a steady stream of articles and stories that focused on the theme. Entitled "Animal Baiting," it began by observing: "Animal baiting refers to the taking of certain animals' lives by others in an agonizing manner."[326] It provided examples of baiting and hounding competitions that had taken place in both Moscow and Petersburg in the mid to late 1880s, focusing on the pitiable conditions of the caged animals and their plight when released before borzois and (less commonly) birds of prey. It noted that a principal organizer of such events was the Society for the Advancement of the Field Qualities of Hunting Dogs. It then summarized a recent article published by an "Old Hunter," who complained that the public did not appreciate the qualities that were tested in these events, which required very different abilities on a dog's part than would actual hunts. If this was so, the piece concluded, why could the wealthy individuals and hunters' societies that sponsored such events not put their money and time to better use by traveling to those regions where peasants and their livestock were actually under threat from wolves and hunt those wolves instead?

On June 8, 1894, a member of the RSPA named A. N. Kremlev gave an impassioned and lengthy speech criticizing the institution of animal baiting at a meeting of the Society's governing board. Noting that these spectacles typically took place every fall and spring, Kremlev quoted from a recent eyewitness account of the hounding of both wolves and rabbits by A. Faresov that had been published in the same St. Petersburg newspaper, *Novoe vremia*, in which Chertkov's article "A Wicked Pastime" had appeared. The rabbits, in Faresov's rendition, cried like children as they were dismembered and in some cases mistakenly attempted to flee into the audience of spectators, vainly hoping to find refuge among them. Kremlev quoted at length Faresov's description of borzois encountering an especially strong wolf:

> This was the struggle of a powerful and clever being, who clearly understood the malice of his enemies, and who would have avoided the fight if it had been possible to withdraw or to gain the sympathy of his enemy by submitting. The bloodied dogs were thrown about the ground as they gripped the wolf, and he bent over lower and lower, his head moved more and more weakly, while new dogs fell upon him with new and fresh strength.[327]

Kremlev informed his listeners that after the dogs had rent the wolf's flesh he was returned to his cage in order to mend so that he would be ready for use in future competitions. In the remainder of the speech he argued that such spectacles should be banned, along with borzoi hunting itself, as it represented a travesty against Christian faith and simple humanity. He compared the fate of the animals hounded by the borzois with those of Christians exposed to predators in the Roman coliseums, and noted that the ticket prices (fifteen, three, or one ruble seats were available) allowed a broad spectrum of people to attend to their moral detriment.[328] He also reminded the audience that their European compatriots viewed borzoi hunting as antithetical to civilized norms by quoting the proceedings of an international humane society conference that had taken place in Zurich in 1869, then expanding on the significance of its message:

> "Hunting with borzois is objectionable on both religious and ethical grounds. Societies for the protection of animals must endeavor with all their forces to disseminate this thought widely by means of the press." And if hunting with borzois is unethical, then a bloody repetition of this hunt is even more unethical. As if it's not enough that people who have nothing else to do hunt animals with hounds, in addition they also prepare for this hunt by means of the baiting and torturing of animals that have been captured especially for this purpose and doomed to lifelong torment; and they do this publicly, inviting everyone to take pleasure in this spectacle.[329]

In addition to such articles and speeches, the RSPA also published fiction that criticized wolf hounding. The November 1905 issue of its journal, now renamed *Defense of Animals*, contains one of the most striking examples of this—a twelve-page "sketch" entitled "A Cruel Pastime: Hounding" by S. Pospelov.[330] At the outset of the story the narrator describes preparations for an unusual summertime competition of borzois sponsored by "The Society for the Eradication of Animals" for which hundreds of rabbits and wolves have been requested from all corners of Russia. Unfortunately, nearly half of the 400 or so rabbits arrive dead or with broken legs, while very few wolves are to be had due to the season. The general who is in charge of the event forces his old servant Finogen, who has been put in charge of procuring the animals, to hand over his own pet wolf, whom he adopted as a pup, for use in the competition. Pospelov's story recounts in vivid and gruesome detail the dismembering of the hares and rabbits by the borzois, which is accompanied by their childlike and frantic screams as they attempt to escape. The servant's pet wolf disappoints the spectators and outrages a visiting count, whose borzois are competing, as he immediately runs over to cower at the feet of his master, who protects him from the advancing dogs. The story is based on a series of emotional climaxes in which pathos mixes with revulsion, as is evident in this vivid description of a female wolf that has been mauled by a pack of ten borzois, which served as the epigraph for this chapter:

> The she-wolf breathed heavily. Her torn and wounded flanks, on which the red meat could be seen, quivered like a pulse. Her tongue, sticking out

from her bloody muzzle, lay on the ground and red froth flowed from it. But, strangely, her eyes did not hold an expression of fierce rage; large and wide, they sooner expressed grief and bewilderment. She looked at the people with a sort of surprise, as if asking why there were people nearby, and not understanding what had happened.[331]

Pospelov's attempt to convey the female wolf's perspective, written in the third person, avoids explicit anthropomorphism but nevertheless utilizes the abused wolf as a window into humanity. Her bewildered and grief-stricken gaze, which contrasts so sharply with the "savage rage" that Chertkov witnessed in the eyes of a wounded wolf as he beat it to death with a stick, represents an interpretive act on the part of the narrator. In each case, the human observers are the ones who impute meaning (or choose to ignore) the possible significance of the animal's gaze. In Pospelov's account the story's narrator uses this juxtaposition to highlight the indifference and callousness of most of those watching, while implying that the proper attitude toward the abused wolf is one of sympathy; for enlightened readers, the tortured wolf becomes a symbol of human cruelty and her gaze a mirror that elicits introspection.

A year after publication of "A Wicked Pastime" Pospelov published another story about wolves entitled "Two Brothers" in a collection called *Tales of Wild Animals*.[332] Unlike his earlier story, the twenty-page third-person narrative focuses closely on the perspectives of two wolf pups sold to a landowner by some peasants, who remark that they would have shot the mother at the den site had she not run off into a state-owned forest. The pups, whose eyes have not yet opened, are put to suckle on one of the landowner's hounds alongside her own pups. His ten-year-old son Seryozha develops a close bond with the young wolves, who are named Serko and Sedko for the darker and lighter shades of their gray coats. During the fall hunting season, the landowner unsuccessfully tries to utilize the young wolves alongside his hounds to chase rabbits, but they do not train well. Later in the winter and increasingly the following spring, they begin to reveal problematic traits: they are unsociable, they disappear from the yard for days at a time, and soon they begin to kill the household fowl. Finally, they maul a sheep together on the grounds. As a result, they are chained. Only Seryozha continues to visit them with tidbits and they trust only him.

At this point, the narrative viewpoint shifts squarely to focusing on the two wolf brothers. Serko manages to escape but is hampered by the chain that still hangs from his neck. Unable to catch prey he begins to starve until one morning hounds are unleashed in the forest where he has taken refuge. They turn out to be the very hounds among whom he grew up, including the female who suckled him. At the end of the hunt, he is cornered and killed by a hunter's dagger thrust: "Something flashed above the head of the wolf, and Serko rolled over dead."[333] The narrative perspective then shifts back to Sedko, who remains chained and pines after his missing brother. Seryozha's father decides to utilize the second wolf for a hounding by his neighbors' borzois. After being deprived of food for two days Sedko is enclosed in a box and released three times before different groups of borzois, who maul him as Seryozha weeps in the manor house. Afterwards he is shut up back in the box, where he dies of his wounds. At the story's end we are told that when Seryozha reached adulthood he sold his father's hunting dogs, dismissed his hunters, and devoted himself to animal husbandry.

Pospelov's story utilizes the young boy as its moral compass. He lavishes affection on the wolves, and they return it in kind. Unfortunately, neither they nor he can prevent the development of their carnivorous instincts as they grow, nor the punitive measures taken by the adults. The restraining chains harden their feelings of isolation and—while the narrative clearly condemns the landowner's treatment of them—it's not clear how the growing wolves could have been kept captive with a positive outcome. As Sabaneev himself had noted in his monograph on wolves, attempts to raise full-blooded wolves to be tame—even if they were captured before their eyes had opened—were rarely successful past the age of six months or so as they would grow increasingly aggressive and uncontrollable after that age.[334]

In the context of his era, Pospelov's stance of identification with the wolves was obviously unusual, but by no means singular. Moreover, while his story privileges the moral sensibilities of a child protagonist, it transcended the domain of children's literature in the eyes of at least some of its readers. A 1906 review of his *Tales of Wild Animals* that appeared in the RSPA's journal, *Defense of Animals*, singled out "Two Brothers" as one of the most successful stories in the collection. The author, E. Chernobaev, provided extended quotes from the *Tales* and analyzed the stories in literary terms, recommending them to both adult and child readers for their artistic merits and their sympathetic portrayals of wild animals, including carnivores like wolves. He emphasized especially Pospelov's ability to create a sense of individuality in his animal protagonists, and compared him to the North American writer Ernest Thompson Seton:

> Each animal that is depicted by the author in a sketch or story appears three-dimensional and draws attention to itself through the gentle manner of the writing not only because of traits of character and degree of

intelligence, but in addition the creature stands before the reader in all its stature, with all its individuality, if one can express it so. And in this sense Pospelov resembles the famous American writer E. Seton-Thompson [sic].[335]

Chernobaev's invocation of Ernest Thompson Seton (1860-1946), who spent most of his childhood in Canada but gained renown as a popular writer of literature about nature and animals for schoolchildren in America, is interesting partly because Seton's most famous work, *Wild Animals I Have Known* (1898)—like Pospelov's *Tales of Wild Animals*—contains a story about the death of a wolf. In addition, Seton was criticized by some of his contemporaries for overly sentimental and anthropomorphic depictions of animals, as Matthew Cartmill has explored.[336]

The intertwined issues of sentimentality and anthropomorphism represent challenges for those who write about animals. Given that any attempt to capture the non-human perspective is fraught, writers must make certain choices. How should the author bridge the gap between animal and human consciousness(es)? How should one represent an animal protagonist's thoughts, given the lack of a shared verbal language? Why are stories about animals so often written for children, and can they successfully appeal to adults? In the remainder of this chapter I will explore four additional authors' attempts to grapple with this complexity in late imperial Russia, then return to some of the larger issues in play at the chapter's end and in the book's conclusion.

The four examples to which we will now turn appeared in 1892, 1895, 1902, and 1907. In 1892 "The Life and Adventures of One Wolf," by an author identified only as A. L., appeared in *Nature and Hunting*. Thirty pages long, it presented the life story of a male wolf from his puppyhood through his maturity and eventual death at the hands of gun hunters. Three years later Chekhov's only story written for children—"Whitebrow"—described a hungry mother wolf's attempts to procure food for her litter, culminating in a startling juxtaposition of the wild and domestic as she encounters a local peasant's puppy. Seven years later, in 1902, a short story by the writer B. K. Zaitsev entitled "Wolves" presented a stylized sketch of the terrible suffering and resulting savagery of a pack of wolves driven to desperation by hunger and fear amid Russia's vast and snow-covered fields during a bleak winter in a compelling emotional tableau. And finally, in 1907 the innovative writer L. D. Zinovieva-Annibal included the story "Wolves" in her collection *The Tragic Menagerie*. Narrated in the first person, it recounts a young girl's traumatic experience of witnessing the suffering of wolves captured by the tsar's huntsmen for use in a wolf hounding competition and her complex sense of identification with the tormented animals. Each of these stories offers a different window into the ways in which wolves were perceived and portrayed amid the shifting sensibilities of late imperial Russia. With the exception of the last, each employs third-person narration along with various degrees of anthropomorphism to attune the reader to the wolf protagonists' perspectives. Among the four, only Chekhov's story was explicitly intended for children. Taken together, they demonstrate both commonalities and differences in the ways in which each writer confronted the challenges of presenting wolves empathetically.

"The Life and Adventures of One Wolf," which is the longest of the four stories, appeared in *Nature and* Hunting in 1892.[337] While the author's identity remains a mystery the narrative—which after all appeared in a hunting journal read by adults—reveals an expert understanding of wolf habitat and behaviors. It also convincingly portrays peasant attitudes toward wolves and accurately describes a number of approaches to hunting them by both peasant and gentry hunters, indicating that the author possessed substantial expertise in these matters. The story is set in a well-known hunting area of that era: the Mikhal'sky Swamp of the province of Riazan, which is located about 150 kilometers southeast of Moscow. The story depicts the depredations carried out by the wolves on the surrounding villages' livestock in ways that closely correspond to the experiences related in memoiristic and other accounts of the day. Yet it focuses equally on human persecution of wolves. Taken together, all of these elements lend the narrative a great deal of verisimilitude, despite the challenges and moments of awkwardness that arise from the author's attempts to convey the perspectives of the wolf protagonists.

"The Life and Adventures of One Wolf" is divided into four sections. Part I describes the young wolf's early life in the remote swampland. Largely impenetrable to humans, the swamp abounds in wildlife and is a paradise both for animals and the few hunters who venture there, as the third-person omniscient narrator recounts. The narrator moves freely between the human and wolf worlds. He writes of the excellent hunting opportunities the swamp contains and its natural splendor, while focusing predominantly on the wolf protagonist's perspective as he traces the daily joys of his childhood discoveries and adventures. Affection reigns within the family group, as the adventurous and curious pups explore their idyllic surroundings. The young wolf and his siblings soon come to prefer the taste of blood and meat to their mother's milk as they encounter prey ranging from frogs, to a duck, to rabbits and eventually piglets and lambs from villages beyond the swamp, which are brought to them by their far-ranging mother, father, aunts, and uncles. As the summer progresses the pups begin to accompany the adult wolves on their nightly hunts and have their first encounters with humans. One night the young wolf and his siblings watch in petrified fascination as the adults kill a bull calf, then escape the shepherds and their dogs with their spoil, and it's clear that such depredations occur frequently. Over the next month the pack of ten wolves attacks sheep, dogs, and once even a horse so that "all the neighboring villages were in upset; not a night passed during which some livestock wasn't killed."[338]

As the wolves' world increasingly infringes on the human one around it, the narrator ventures beyond their ken at strategic junctures, providing a bridge between the wolf and human perspectives. For example, he quotes a conversation between a peasant father and son as they are passing through a nearby meadow and hear the wolves howling. The son suggests they find and follow the wolves' pathways to their den and smother the pups. The father scoffs at his son's naïveté, retorting that they would become mired down and lost in those same paths and recounting tales of the vengeance on villagers undertaken by enraged mother wolves who have lost their young.

Part II of the story depicts the growing hardships that confront the pack with the onset of winter. Their troubles begin when a local landowner with a large kennel of borzois and scent hounds decides the time is right to hunt the pack, whose den site his huntsman has identified in advance. The narrator emphasizes that the landowner waits until late in the season both for the swamp to become less impregnable and for his gentry hunting friends to arrive. While fictional, this corresponds to the prevailing practices we have repeatedly encountered in non-fictional sources and was one source of tension between peasants' and landowners' perspectives. The local peasants, "whose patience has run out," serve as drivers, but only two of the wolves in the pack are taken.[339] The wolves move several versts away to an adjacent but less desirable swampland. There a local peasant lures the wolf and his sister with carrion and shoots her. Shortly thereafter, the wolf protagonist's father and uncle become sick and die after eating a horse laced with poison. The narrator presents the effects of the poison from the viewpoint of the wolves: "[Their death] came about for an unknown reason, near the carcass of a horse. That carcass, apparently, was harmful. The wolves vomited blood, then they turned around and around, and died."[340]

The growing wolf, now preeminent among his siblings and known by them and the local peasants as the "Black One" (*Chernyi*), learns from this and other experiences, such as when his leg is temporarily caught in an old and rusted trap. The narrator generally restricts himself to precise and plausible description of the wolves' movements and activities, as well as their surroundings. Occasionally, however, he skirts into anthropomorphic territory by describing their mutual affection and their growing understanding of the world around them in ways that can seem overly rational. Rarely, he even uses quotes to encapsulate their thoughts. For example, the dark wolf suspects something is amiss as his sister approaches the carrion over which she will be shot: "Something's not right," noted the dark one, directing his nose in all directions and sniffing the air . . . Carrion never smells at all in the frost."[341]

The Black One and his other siblings make it through the winter, then return in the spring to the Mikhal'sky swamp, where their mother has found a new mate. Despite their losses over the winter the newly reconfigured pack with the spring's seven new pups now numbers a dozen. The Black One teaches his mother's new mate to hunt hares, and the older wolf teaches him how to kill domestic animals including geese, sheep, cats, and dogs. The pack assiduously avoids killing animals in the nearest villages, echoing a theme we have seen repeatedly in other sources: "A chicken from the nearest village could have walked between the feet of any of the Mikhal'sky wolves, and no one would've touched it."[342] Unfortunately for the pack, in late summer a local borzoi hunter gathers a hunting party that includes twenty peasants from a nearby village to serve as beaters. He also enlists five as shooters. Guided by a seasoned Siberian huntsman, they set snares and station themselves skilfully and quietly. In the ensuing hunt, every single one of the wolves with the exception of the Dark One is killed by the borzois or guns. The narrator's sympathy for the wolves, despite his copious knowledge of hunting technique and admiration for one of the hounds (about whom he promises to write a separate story), becomes evident as he castigates one of the hunt's main participants, a drunken colonel, in the last sentence of the section: "And the colonel complained so about the fact that they'd let the Black One escape! The bloodthirsty monster wasn't satisfied with eleven wolves . . . "[343]

Part III depicts the full maturity of the Black One, who becomes the dominant wolf in the area and the subject of legend in the surrounding villages. Children are warned that he will eat them, and some of the old peasant women decide that "for sure it's a werewolf."[344] He finds a mate and several pages are devoted to describing a hunt in which three men from the city track the wolves through the snow. This culminates in a fusillade of shots, one of which wounds the Black One in his right foreleg.

Part IV traces the aftermath of this event, emphasizing his mate's solicitude for him during his recovery. Here again the narrator presents the female wolf's thoughts verbally in an anthropomorphic way: "Grief overcame the Golden One, as she saw her handsome one mutilated: '"Oh damned people, crippling animals for fun!"'[345] The Dark One is able to overcome his lameness by developing a new specialty of hunting geese and domestic fowl, although he eventually loses his mate. He also becomes increasingly wary and sophisticated about humans, learning to observe carrion from a distance for several days before approaching it and becoming adept at avoiding steel traps and snares. Hunters return repeatedly in pursuit of the legendary wolf—now utilizing the Pskov technique—but he learns to evade them by avoiding open land without beaters or flags and even by observing the preparations from a distance prior to each hunt. Finally, however, a determined hunting party outwits him by mounting six drives in a single day and again arming some of the beaters. He escapes but has been shot through the jaws. This grievous injury leads to a process of slow starvation. By the story's end he has been reduced to a vestige of his former self. As he lies sleeping in the bushes, dreaming of his early happy life, hunters approach in yet another drive. He runs forth from the bushes, is hit by several gunshots, and dies. In a final declaration of admiration for his wolf hero and condemnation of the people who persecute him, the narrator declaims:

> What an amazing amount of shrewdness, life force, and energy perished with him!
> Yes, he was one of the most impressive examples of wolf kind! And you, pitiful and cruel people, you were able to vanquish him only in this tormented, maimed, and half-alive state . . . [346]

Taken as a whole, the story evinces a deep knowledge of hunting, wolf behavior, and the attitudes of gentry hunters and local peasants toward wolves. It represents the entire life cycle of its wolf protagonist in great detail and with verisimilitude. It also reflects some of the challenges that writers of the day confronted in trying to represent the world in a wolf-centered narrative. While the author's extensive knowledge of wolf's daily activities, means of procuring food, and social lives lend "The Life and Adventures of One Wolf" a great deal of plausibility and interest, there are times where the narration falters amid overly emotive expostulations or jarring attempts to represent the wolves' perspectives verbally. Chekhov, as we shall see, approached some of the same challenges in a different way.

Figure 12 Alexei Stepanovich Stepanov, Wolves on a Winter Night (c. 1910). Image copyright Lebrecht Music & Arts

In the fall of 1894 the editor of *A Children's Reader* (*Detskoe chtenie*), D. I. Tikhomirov, approached Chekhov with a request that he craft a submission for the journal. Chekhov originally responded that it would be difficult to write specifically for children. Nevertheless, he complied, submitting the story "Whitebrow" in April 1895.[347] The story appeared in *A Children's Reader* in November 1895 and subsequently in an 1899 collection of stories for children compiled by M. Vasil'iev and entitled *Tales of Life and Nature by Russian Writers*, as well as separately in brochure form (against Chekhov's wishes due to copyright concerns). It remained the only work he wrote specifically for children and was incorporated into the first edition of his collected works edited by A. F. Marks. Despite its intended audience, the story—as the early critic I. Belousov noted in 1899—maintains Chekhov's characteristic style: although "the story's plot is simple and straightforward, its elaboration is 'Chekhovian.'"[348]

"Whitebrow" begins with a third-person description of a hungry and timorous mother wolf leaving her den on a frigid March night to seek food for her three pups. Her hunting is thus linked directly with her maternal instincts:

> The old she-wolf was in poor health and wary; she recoiled from the slightest noise and worried the whole time about her cubs being harmed by someone in her absence. The scent of traces of humans and horses, stumps, the piled-up firewood, and the dark dung-strewn road frightened her. It seemed to her that beyond the trees in the dark people were standing, and dogs were howling somewhere beyond the forest.[349]

The old wolf's inadequate sense of smell causes her to confuse the scent of fox and dog or even to lose scent trails altogether. Handicapped by her age and infirmity, we learn, she is unable to pursue the calves and sheep that she formerly hunted and must usually settle for carrion. The she-wolf turns out to be heading for an isolated cabin four versts from her den in which a seventy-year-old retired machine worker named Ignaty lives with his three dogs. Chekhov's third-person narrator encapsulates Ignaty's character in a humorous vein that alludes to his apparent fondness for alcohol in a way that both a child reader and the old wolf might perceive it: "Sometimes he sang and while doing this staggered heavily and often fell (the wolf thought this was from the wind), and shouted: "He's come off the rails!"[350]

Hoping to find a lamb in Ignaty's shed, the wolf climbs in through a gap in the thatch roof. Amid the resulting commotion, she snatches the first smallish warm thing she encounters and runs off with it in her mouth as Ignaty's large guard dog, Arapka, barks. At this point Chekhov's story takes an unusual twist that places it in contrast with almost all of the stereotypes and understandings of wolves that prevailed in the Russia of his day and that directly call into question the stark division between the wild and the domestic that underlay them.

While we have repeatedly encountered accounts of wolves attacking and eating village dogs, the old she-wolf reacts with a mixture of discomfort and disgust to the discovery that she has made off not with a lamb but with a large puppy:

> The she-wolf stopped and laid her burden on the snow for a rest and to begin to eat it, and suddenly jumped back with disgust. It wasn't a lamb,

> but a puppy—black with a large forehead and long legs, a large breed, with the same sort of white patch on its whole forehead as Arapka had. Judging by its behavior he was stupid, a simple yard dog. He licked his bruised, wounded back and, as though nothing was the matter, wagged his tail and began to bark at the she-wolf. She growled like a dog and ran away from him. He followed her.[351]

The puppy follows her back to her den, where it spends the day playing with her three pups as she hungrily gnaws on an old bone. This portion of the story gains narrative tension from the reader's sense that the she-wolf may be driven from hunger to eat the puppy, whose stupidity Chekhov repeatedly emphasizes. And indeed, as her own pups feed on her milk, she resolves to do so:

> The wolf pups suckled, and the puppy, who wanted to eat, ran around in a circle and smelled the snow.
> "I'll eat him now . . ." the she-wolf decided.
> She approached him, but he licked her on her muzzle and began to whine, thinking that she wanted to play with him. In the past she had eaten dog, but the puppy smelled very doggy, and with her poor health, she could no longer stand the smell. She felt disgusted and moved straight away.[352]

Chekhov's presentation of her attitude toward the puppy is nuanced in the sense that she doesn't explicitly feel compassion or kinship for it. Rather, her revulsion at the idea of eating it reflects her own delicate health and its unpalatability. At the same time, her lack of aggression toward the puppy as it spends the day playing with her own litter creates a strong contrast with prevailing stereotypes about wolf behavior and bloodthirstiness and perhaps even implies some degree of empathy for the puppy on her part. In addition, as in "The Life and Adventures of One Wolf," her thoughts are encapsulated via brief quotation in a few instances, primarily relating to her feelings of hunger, her immediate surroundings or her actions, as in the passage above. Yet Chekhov carries this off in a less jarring way than the earlier author, perhaps because he presents only very basic ideas—"I'll eat him now"—in the form of verbal quotes and because of his understated and straightforward style of narration.

Ultimately, the puppy runs off homeward. When the she-wolf goes hunting again that evening, however, and returns to Ignaty's shed in search of the lamb, Whitebrow follows her into the warmth inside and his excited barking arouses the guard dog Arapka and Ignaty; again, she must retreat with nothing. The story ends with Ignaty, who is convinced that it was Whitebrow who twice dug through the thatch roof, berating the puppy, unaware that a wolf has been near in a typically Chekhovian mixture of miscomprehension, humor and pathos.

"Whitebrow" presents a sympathetic and compelling portrait of the old wolf. While written for children, the story—as Belousov had noted in 1899—is well-crafted and consistent with Chekhov's mature style. It delicately attunes the reader to the mother wolf's outlook and moods, which are linked with her hunger, ailing health, concern for her pups, and fear of humans. It juxtaposes her knowledge of suffering with the naïveté of Whitebrow and explores the interplay between their two worldviews in a series of creative juxtapositions. It delves into her thoughts—even utilizing the occasional brief quotation—but does so primarily via third-person narration unobtrusively and convincingly by anchoring them in her immediate perceptions of her surroundings and her physical senses of sight, hearing, taste, touch and smell. Chekhov's creative juxtaposition of wolf and puppy, one wild and the other domestic, succeeds in recasting the relationship between the two in highly original terms.

B. K. Zaitsev (1881-1972) was a writer of prose, a playwright, and a translator.[353] The son of an engineer and of gentry background, he spent his childhood in the province of Kaluga about 200 kilometers southwest of Moscow and was an avid hunter during his youth. He met Chekhov in Yalta at the age of seventeen and considered him a model. His short story "Wolves," published in 1902, helped to establish his reputation. Over the following decades he published several collections of short stories prior to emigrating from Russia in 1922 and eventually settling in France. Looking back at his early writings in 1916, Zaitsev remarked that naturalism and impressionism were important elements of his approach.[354] In the commentary that accompanies her compilation of nineteenth-century Russian hunting tales, M. M. Odesskaya notes: "Loneliness, death, the indifference of nature—this is the thematic core of his stories."[355]

These qualities are strikingly evident in Zaitsev's short story "Wolves."[356] The narrative is divided into four brief sections of about one page each. Zaitsev's compressed and symbolic language utilizes an almost poetic repetition of key imagery and motifs that impact the reader powerfully and viscerally. We are immediately immersed in the desperate plight of a pack of starving and persecuted wolves amid Russia's unforgiving winter landscape:

> It had stretched on for a week already. Almost every day they besieged and shot at them. Withered, with ragged sides from under which the ribs protruded ominously, with glazed eyes, resembling specters of some

sort in the cold white fields, they crept without discrimination or direction when they were roused from their beds and senselessly rushed and wandered over the very same places again. And the hunters shot at them confidently and accurately. During the day they bedded miserably in little bushes that offered at least a bit of cover, hiccupped from hunger, and licked their wounds. At night they clustered in small groups and wandered in single file across the vast empty fields. The dark malicious sky hung over the white snows, and they dragged themselves toward that sky, but it ceaselessly ran from them, and everything was just as distant and gloomy.[357]

In these opening sentences Zaitsev creates the mood that will reign over the story through language that is stark, harsh, and deeply unsettling. He portrays the wolves as emblems of suffering at the hands of both the hunters and the ferocious cold of the Russian winter. A malevolent heaven looms over them, as they endlessly and hopelessly seek escape from an existence of pure suffering. Unable to find shelter, food, or protection from the hunters' bullets they can only howl in desperation: "their howling, exhausted and anguished, crept over the fields, and froze within a verst or half more, and didn't have enough force to fly upward to the heavens and echo in a shriek about cold, wounds, and hunger."[358] In the final paragraph of the first section, a female engineer who is returning to her quarters at a nearby coal mine substation, hears their distant howls and crawls into her bed under the sheets muttering: "damned, damned."[359]

Section II (each is identified by a roman numeral, almost as grouped stanzas might be in a long narrative poem) focuses on one group of the wolves as they follow in the tracks of their leader, an old gloomy male, limping from a bullet in his leg, each thinking only of survival and the discomfort of their paws being cut by the sharp snow. They scatter and regroup amid the gusts of wind and snow, until one of them—a gaunt younger male—refuses to continue. Zaitsev verbalizes the stricken wolf's thoughts in the form of speech:

"I won't go further" he said brokenly and gnashed his teeth. "I won't go, it's white around, . . . white all around . . . snow. It's death. Death."
    And he pressed close down to the snow, as if listening.
"Listen . . . it's speaking."[360]

The other wolves leave him and continue their endless trek. They see the outlines of a village and smell the livestock. The old lame wolf, however, forbids them from going there. Section II ends in a reprise of the same phrase voiced by the female engineer at the end of Section I, but this time snarled by two of the wolves as they cast anguished glances at the forbidden village: "damned, damned."[361]

Section III singles out two of the wolves in the pack, who question the older lame wolf's leadership. This shift toward their perspectives is accompanied by a further personification of the hostile world that surrounds them, which is again associated with a malevolent force in the heavens: "The lifeless snows peered at them with their pale eyes, dimly reflecting something from up above . . . "[362] They sense that the wolf who gave up and stayed behind to die is right, that all of nature seeks their death, "that the white emptiness truly hated them, hated them because they were alive . . . Despair overcame them."[363] Their anguish leads to fights: "There were several more scuffles: brutal, unnecessary, and unpleasant."[364] Another pair of wolves lie down to await death but are so overcome by the horror of being alone that they run to catch up to the others.

Section IV brings the dramatic tension to a close. As the wolves' desperation reaches a climax the lame old leader of the pack, realizing his life is in danger, speaks: "Comrades," said the old wolf. "All around us are fields that are enormous, there's no way to get out of them quickly. Do you really think I'll lead you and myself to doom? It's true I don't know exactly where we should go. But who does know?"[365] The pack's response is immediate and as unforgiving as the bleak environment on all sides; they close in, attack, and devour him:

> Dozens of teeth, all just as sharp and burning, as [the first], bit into him, rent, tore out his insides and ripped off pieces of fur. They all churned into one ball that rolled over the ground, they all nipped whatever they felt pressing against their teeth. The ball snarled, for a moment eyes would glisten, teeth and bloodstained jaws would show briefly. The malice and anguish that had snuck forth from these ragged and thin bodies hung over this vengeance in a choking cloud, and even the wind couldn't drive it away. [366]

After ten minutes, all that is left of the old wolf is his head and scattered bones. The rest of the pack, blood-spattered and spent, drift apart to separate beds in the snow as the blood droplets turn to ice in their fur. They begin to howl, but each now howls alone rather than in a chorus as the story ends: "Nothing could be seen in the dark, and it seemed as if the very fields were groaning."[367]

Zaitsev's story transcends the strictures of realism to present the world of the wolves symbolically in an almost poetically dense narrative. The emotional texture of their tortured existence correlates with physical sensations of hunger, pain and cold and a metaphysical sense of persecution. The distant hunters whose bullets hit them so accurately, causing such pain, create a continual sense of dread. The malevolent forces of nature embodied in the wind, snow, and cold are equally or more fearsome than these human nemeses. The wolves are overcome by angst, rightly convinced that the world around and above them is directed at their extinction. The verbal interchanges that precede the story's violent climax would seem contrived in another narrative context. In the setting of Zaitsev's story, though, they seem almost like abstract representations of speech, as if the narrator is interpreting the wolves' fearful and fateful interactions for the reader alongside his portrayal of the hostile world that surrounds them. The pack's frenzied killing of the old wolf, horrifying and primitive, compresses the fundamental violence of their lives into a stormy and snarling ball of teeth, blood, and eyes rolling across the snow amid the darkness. Their solitary howling at the story's end, alone amid the fields that themselves seem to be groaning, culminates the story's dominating theme of existential isolation.[368]

Figure 13 Alfred Wierusz-Kowalski (1849-1915), A Wolf Pack (unknown date). Art Collection 4/Alamy Stock Photo

The fourth story I will treat is the 1907 first-person narrative "Wolves" by L. D. Zinovieva-Annibal (1866-1907). Zinovieva-Annibal, remarried to the famed symbolist Vyacheslav Ivanov and mistress of their influential literary salon, became an accomplished writer in her own right. "Wolves" appeared in her pseudo-autobiographical collection *The Tragic Menagerie (Tragichesky zverinets)*, which was published the year she died and—after decades of neglect during the Soviet period—is now recognized as her most significant work. The twenty-page story is one of three in the collection that focuses on the young narrator Vera's interactions with the wild animals (tamed bear cubs, a wild crane, and the wolves) that live near her noble family's country estate, which is located about 100 versts from Petersburg near the Baltic Sea. In each case she unwittingly perpetrates or witnesses cruelty to animals with which she also deeply identifies. In one story, for example, she forgets to feed her pet crane for several days and finds it dead due to her neglect. Jane Costlow, who has both translated and written eloquently about *The Tragic Menagerie*, observes that it "recounts both childhood and the genesis of adult sensibility."[369] The stories that focus on animals, in Costlow's view, ask the reader "to meditate on the fate of animals, but also on the fate of the animal within the human, on the intersections of nature and culture, the wild and the tame."[370] From my perspective, "Wolves" is invaluable for the ways in which it both recapitulates some of the central features of the cultural discourse around wolves that we have explored and reinscribes them in the emotional landscape of a precocious female child on the verge of adolescence. It culminates late imperial era literary portrayals of wolves in a strikingly evocative and powerful way.

"Wolves" begins with Vera's excitement late one autumn at the news that the "tsar's hunt" has come to their village to procure wolves from the nearby forest for later use in wolf hounding competitions at the tsar's park in the capital. The head huntsman and a nobleman who accompanies him, Vladimir Nikolaevich, stay over at her family's house and the latter explains in detail how the hunt will be carried out:

> The huntsmen cordon off one part of the forest with high, strong nets; in all other directions local peasants, called in by the hundreds, are closely interspersed. The peasants are armed with clubs and rakes, and yell loudly, their yells keeping the wolves from getting by them out of the forest. Huntsmen on horseback make their way into the woods with packs of hounds. The hounds smell out the wolf and pursue him toward the net with their high-pitched barking; the wolf hits it at a run, then a second net

falls from above; he thrashes and gets so tangled there's no getting out. The huntsmen arrive. They bend the animal's neck to the ground with a two-pronged clamp; they tie his legs like a sheep's; once they've turned him on his back, they put a thick short stick crossways into his open mouth, and when it's tightened with a rope at the withers, they lift the animal by its bound legs onto a thick pole. Once they've hoisted the pole onto their shoulders, two men carry the wolf face upward onto the main road, where huge covered wagons, like freight cars, await the captives.[371]

Vladimir Nikolaevich adds that they break one of each wolf's legs so that it can't run away too quickly during the subsequent hounding and "also so they can't attack."[372] Initially, the only listener to express discomfort with this mode of entrapment and its aftermath is Vera's English governess, Miss Florry, who labels it "barbarian cruelty."[373] Vera herself is too preoccupied with a girlish infatuation for Vladimir Nikolaevich and excitement at witnessing the unusual spectacle to worry about the wolves. The upcoming hunt, as recounted by the nobleman, echoes descriptions of the gun hunts organized by The Imperial Moscow Hunting Society that we encountered in Chapter 2. The use of nets rather than guns, however, allows for the wolves to be captured instead of being killed, and will turn out to be thematically crucial for Zinovieva-Annibal's literary portrayal.[374] I have not encountered the method of breaking a wolf's leg prior to a hounding in historical sources: while such a practice may have existed, Zinovieva-Annibal may also have included it in this precise account of the hunt as a way of engendering Vera's (and by extension the reader's) sympathy for the wolves, which would be consistent with other aspects of the narrative.

On the morning of the hunt Vera sets off with her governess, the peasant coachman Fyodor, and other members of the household before dawn to observe. After they have arrived at the forest, Vera wanders among the trees and imagines herself as a wild nomadic princess responsible for protecting her people by leading them to safety through the ogres and other enemies that encircle their encampment. In her imagination she conflates Russian culture's apprehensions about rabid wolves, which she has internalized, with the upcoming hunt:

> But the wolves ... All the wolves in the forest have turned rabid ... What can be more terrible than a rabid wolf? He fears no one, throws himself at a crowd, and bites now one and then another ... and they turn rabid, too ... You have to tie them up ... It's the enemy who has infected the wolves with rabies, so they'll destroy my people, but we've hung nets on the trees

to protect ourselves from the wolves, and now I'm keeping watch while all my people sleep.³⁷⁵

Overcome with horror and fright at her vision, Vera runs back to her governess "howling like a wolf in horror at being pursued."³⁷⁶ From this point forward, the narrative emphasises this duality of her fear and simultaneous identification with the persecuted wolves. She argues with herself as she witnesses the bound wolves being carried past and put into cages with iron bars. Her inner dissonance reflects the struggle between yielding to the compassion that the sight of an animal in pain arouses and her understanding that wolves are predators along with stereotypical notions of what this means:

> I feel sorry for the wolves. That repulsive, slippery, flabby feeling comes creeping into my chest. I shove it away: wolves are evil, they eat sheep, they ate my donkey . . . Wolves are evil and disgusting cowards! They attack loners in packs. . . . What nasty eyes! . . . The little eyes watch with evil horror, like little coals—of course, like searing little coals! At night they glow like little green lanterns, wolf eyes do.³⁷⁷

As she struggles to disentangle her conflicted feelings, Vera focuses increasingly on the wolves' eyes. At first, she pictures them in ways that reflects popular conceptions we've encountered—like coals or green lanterns that light the rapacious predators' way in cowardly night-time attacks on defenceless domestic animals. Soon, however, a jarringly direct encounter with one wolf, distinguished from the others by a terrible wound in his side, forces a rupture in her thinking as compassion overcomes her fear and culturally ingrained hostility:

> But I've had a good look, and I've long since been crying. This wolf has been poked through the side with a rake. He breathes through a hole in his side. The air hisses, and it seems as though I hear it through the hole; the edges of his wound move up and down. It's horrible. The wolf's teeth have bit the stick in his mouth; quite close to my face, where it's pressed against the bars, are his eyes. In their corners I see the white part. It's all bloody. His pupils strain straight into my pupils. Unbearable pain, furious hatred, and sorrow are condensed in them, along with a final, hopeless, settled horror. These pupils have laid a spell on me and I, like him, clench my bared teeth and drill my wild pupils, dried now of their recent tears. I hear my grimace. The skin stretches dry. With my ears I hear my repellent

> wolf face, the hatred, horror, and pain in its pupils and in its stretched-out lips ... The air keeps hissing as it bursts from the bloody hole in the side, and the sides of the wound flap up and down with rapid, feverish breaths. How horribly the body is made! If you poke it through there's a kind of bloody softness, and then there something separate—the liver? the heart? a lung? What is that naked bloody thing that lies open in the living body of the wolf? Why doesn't he howl? Why doesn't he yelp or howl?[378]

In this startlingly intimate and painful moment, Vera recognizes that both she and the wounded wolf share a common mortality rooted in their animal nature and their very flesh. Their interlocked gaze precipitates a cataclysm of emotions within her. His pupils—in which she reads pain, hatred, sorrow, and horror—exert a "spell" on her, transforming her psychically into a wolf herself. This extraordinary encounter melds notions of the supernatural with her visceral, indeed organic, sense of identification with the wolf; she sees herself in him, and the wolf in herself. As the cart sets off, and she imagines the wounded creature's suffering all along the jolting road to the capital, she voices a "frenzied, animal howl" and runs straight into one of the nets set for the wolves, completing the identification: "The net had me entangled. Then I was seized with rabid horror, and started to lash out, bellowing and whooping, kicking, thrashing my arms, biting."[379]

In the aftermath of this traumatic experience Vera speaks first with the coachman Fyodor and then her ailing mother about her feelings of compassion for the wolf and sorrow over its fate. Fyodor reassures her that animals do not fear death in the same way as humans because they are without sin, unlike humans. Her mother, herself suffering from a debilitating illness, remarks provocatively: "Is it really all that dreadful to suffer? It's more dreadful to watch and feel pity."[380] Both adults' perspectives, grounded in such different worldviews, overlap in their emphasis on privileging the human observer as the site of true ethical, moral and even emotional significance. The human is the one who grapples with the meaning of transgression, suffering and death while the animal simply endures its fate.[381]

Each of the four stories I have presented confronts the challenges of representing wolves' perspectives and evoking empathy because of the hardships they face in different ways. "The Life and Adventures of One Wolf" benefits from its author's expert knowledge of wolf behavior to present a compelling "biography" of its protagonist from his birth to his death at the hands of gun hunters. Chekhov's "Whitebrow" utilizes an understated style of narration rooted in physical sensations and immediate perceptions to convey the moods of the aging she-wolf as she worries over the fate of her litter and spares the life of the unexpected and unwelcome puppy. Zaitsev's haunting tale employs penetrating imagery to create a sense of overwhelming and all-encompassing dread amid a densely symbolic landscape of darkness, cold and death. Zinovieva-Annibal leverages the intensity of a young girl's perceptions to highlight the cruelty of an outdated cultural institution that torments wolves for the tsarist aristocracy's amusement.

While Zinovieva-Annibal's first-person story gives voice to the wolves through the mediation of Vera's highly developed capacity for empathy and blurred sense of identity, the three third-person narratives directly verbalize the wolves' thoughts in one or more instances, although in varying ways. In a 2015 book entitled *Talking Animals in Children's Fiction: A Critical Study*, Catherine Elick argues that providing animals with a voice also provides them with agency. Drawing on M. M. Bakhtin's theory of dialogism, she asserts: "readers, when they engage with animal characters who can express and define themselves through human language, enter into a dialogic relationship with animals that stimulates ethical interactions with them."[382] In other words, by investing animals with language we are ascribing human-like emotions and attributes to them, and correspondingly assuming moral obligations in our relations with them. Elick's use of Bakhtin is helpful in explicating the effects of verbalized thinking in stories like these. Each verbalizes the unspoken thoughts or actual utterances of its wolf protagonists sparingly, but the attribution of an ability to think verbally lends them a moral stature and right to empathy in our eyes.

In Zinovieva-Annibal's first-person narrative, the interlocked gaze of her protagonist Vera and the wounded wolf—rather than verbal utterances attributed to the wolves—provides the fulcrum for the story's overarching moral themes and heightens its emotional impact. The power of the gaze to elicit such a reaction from the human observer allows for a different sort of anthropomorphism than that linked with speech, or perhaps more accurately for an effacing of the boundary between the human and animal consciousnesses. Vera's compulsion to stare at the stricken wolf, and the reader's inability to wrench our attention away from her gaze, demonstrate the power that such literary portrayals had to effect, as well as reflect, changes of attitude towards the feared predator. While each of these literary works approaches the challenge of surmounting the divide between human and animal consciousnesses differently, they demonstrate a common desire to represent wolves more sympathetically. In the conclusion I will delve further into the ways in which we can interpret moments like these, particularly those involving the gaze, then try to ponder some of the attitudes towards wolves which we have explored from the vantage point of the present.

Conclusion

The passage that I've come across most frequently in perusing wolf literature is Aldo Leopold's description of a dying wolf's eyes in "Thinking like a Mountain," one of the briefer essays in his posthumous 1949 collection *A Sand County Almanac*. Leopold (1887-1948) was one of the pioneers of twentieth-century ecology and chose to anchor an epiphany on his path to developing a less exploitative attitude toward nature (a "land ethic") in the same interlocked gaze of hunter and vanquished wolf that we have encountered repeatedly in the very different context of nineteenth-century Russia. He tells of shooting a female wolf and one of her pups while working as a forest ranger in the American southwest in the 1910s and early 1920s, where one of his duties was to participate in wolf eradication programs sponsored by the United States government:

> We reached the old wolf in time to watch a fierce green fire dying in her eyes. I realized then, and have known ever since, that there was something new to me in those eyes—something known only to her and the mountain. I was young then, and full of trigger-itch; I thought that because fewer wolves meant more deer, that no wolves would mean hunters' paradise. But after seeing the green fire die, I sensed that neither wolf nor mountain agreed with such a view.[383]

In a close reading of this scene Andrew C. Isenberg remarks that one should interpret it as a parable rather than a real-time description of an epiphany, adding that the true metamorphosis in Leopold's worldview took place more gradually and later in his career after he had joined the Department of Game Management at the University of Wisconsin at Madison.[384] He also argues that Leopold's recognition of the importance of wolves in an ecosystem—the significance of apex predators in modern ecological parlance—relied on earlier shifts in American cultural attitudes toward wolves.[385] He singles out Ernest Thompson Seton's 1898 portrayal in *Wild Animals I Have Known* of the "outlaw" wolf Lobo, whom Seton managed to kill by exploiting

Lobo's concern for his mate, as one of the literary moments that set the stage for Leopold's later evolution. Seton's emphasis on Lobo's human-like qualities in his anthropomorphic portrayal and his attribution of moral standing to animals, he asserts, was one of the cultural moments that helped to prepare the ground for Leopold's later recognition of wolves' paramount significance in the "moral ecology" of the natural world.[386] This, in turn, helped to set the stage for a larger reappraisal of the predator eradication programs sponsored by the US government in which both men had participated during successive eras, which led to the virtual disappearance of wolves in the lower 48 states by 1950.[387] Isenberg is, of course, utilizing both Seton and Leopold as markers for broader changes in society's attitudes toward wolves, but his tracing of the interrelationship between their stances and the larger American context holds illustrative parallels for the evolution of attitudes toward wolves in imperial Russia as well.

In considering Russians' evolving attitudes toward wolves during the late imperial period, I've also found those moments in which the human gaze probes or interlocks with that of the wolf to be among the most meaningful, evocative and compelling. We have encountered this trope repeatedly across the decades in a range of texts and a variety of situations. We first witnessed it when Princess Boriatinsky described the "utterly red" and fearsome eyes of the wolf her husband had vanquished with the help of his borzoi wolfhound, Beast.[388] For the princess, the wolf epitomized the ultimate threat to the hierarchical world of privilege and patriarchy over which her husband presided, a living embodiment of untamed nature bound and rendered powerless and mute by the prince in his fierce protection of their peasants.

The mature wolf that Nikolai Rostov's borzois pinned down and Danilo subdued and trussed alive to a "shying, snorting" horse represented an extension of the same qualities. But as readers of Tolstoy's novel we, alongside Nikolai, had witnessed the wolf in his original state of wildness—believing that he was unobserved—as well as during his subsequent ferocious struggle with the Rostovs' pack of borzois and, finally, at the moment of his capture. Tolstoy provided us with a sense of the texture and depth of the wolf's life experience, and we sensed his visceral shock at first feeling Nikolai's rapt eyes on him and later the magnitude of his loss as the hunters prodded him after the hunt and peered into his "wide, glassy eyes ... [which] looked at them wildly and at the same time simply."[389]

We next stared into the eyes of a wolf together with Chekhov's gentleman hunter Nilov as he struggled upright and face-to-face with a rabid wolf.

Yet this time human and wolf were linked by their shared susceptibility to the dread disease of rabies and Nilov, despite his acute awareness that the crazed predator could transmit the terrible illness to him, perceived not evil but suffering in the wolf's eyes, witnessed the animal's tears, noticed their resemblance to human ones, and wondered: "What did the sick wolf feel?"[390] These stirrings of empathy corresponded to Chekhov's more sympathetic portrayals of wolves in his journalistic and other fictional accounts as well.

Five years later Tolstoy's collaborator Chertkov peered into a dying wolf's eyes in the encounter that bears the greatest superficial resemblance to Leopold's. Like Leopold, he ruminated on the moment during which he stared down into the eyes of the wolf he had shot as it expired. And like Leopold, he experienced a revulsion of feeling at his act of bloodshed and anchored an epiphany in the moment of observing the wolf as he wrote about it years afterwards. But Chertkov's article "A Wicked Pastime" focused primarily on his own personal moral awakening and his call to other hunters to renounce blood sport rather than the rise of the broad ecological worldview that Leopold would moor in so similar an experience sixty years later. Moreover, unlike Chekhov's gentleman hunter, Chertkov perceived the same savage rage in the eyes of the wolf he was beating that had characterized earlier accounts of wolves, rather than the glimmerings of something akin to humanity sensed by Nilov. The wolf remained an emblem of ferocity, its eyes signifying hatred and otherness, and Chertkov's confession centered on his own spiritual metamorphosis as a human who observed and differentiated himself from the beast.

Pospelov's 1905 depiction hearkened back to Chekhov's varying portrayals of wolves and especially his description of a Moscow wolf hounding open to the public. Pospelov's representation of the "grief and bewilderment" in the "large and wide" pupils of the she-wolf so mistreated and mutilated in a public baiting emphasized that "her eyes did not hold an expression of fierce rage."[391] He consciously employed the trope to reject the tradition, to which Chertkov had cleaved even while renouncing the acts of hunting and killing, of depicting wolves as savage. Rather than offering a window into the hostile enmity of a vicious animal, the eyes of Pospelov's stricken wolf provided a mirror that reflected the sadism of the event's organizers and the spectators who had paid to witness her suffering—a reflection of Russian society that his readers were meant to heed.

Zinovieva-Annibal's portrayal of a wounded wolf's eyes through the perceptions of a precocious girl in her first-person narrative "Wolves" provided the most piercingly intimate and distressing exchange of a gaze between human and animal that we have encountered. Vera's initial horror leads her first to identify with the stricken wolf through an overwhelming psychic metamorphosis and then—in a form of psychological self-defense—to attempt to rationalize its fate in her conversations with others. At the end of the story the reader is left unsure of her final stance on the wolf, leaving us with a sense of ambiguity that perhaps corresponds most closely among these various accounts to our modern sensibilities.

In his 2011 essay "The Gaze of Animals," Phillip Armstrong synthesizes the work of Berger, J. Derrida and others to provide an overview of changing interpretations of the human–animal gaze over time. In the course of doing so he touches on the ways in which humans have perceived the gaze of wolves as well as other real and phantasmagorical predators such as dragons. He notes that classical and medieval sources often portrayed the eyes of nocturnal predators including wolves as actively emitting light, which they associated with fire or supernatural forces that could bring harm to the humans transfixed by their stare (the "evil eye"): "Conceived as a current, flame, fire, stream of particles, or corporeal ray, eyesight was not just an active force in itself but also a vehicle for other physical effects: poisons, contagions, influences of various kinds."[392] With the rise of modern anatomy and biology, scientists demonstrated that the glowing eyes of predators at night had a physiological basis in the reflective *tapetum lucidum*, which enhanced their nocturnal eyesight through reflection. Yet the earlier conceptions continued to live on in folk superstitions and the literature of romanticism. By the late nineteenth and early twentieth centuries, however, writers were representing the animal gaze very differently: "In the fictions of Wells, Kipling, and Conan Doyle we see the fiery animal gaze extinguished by the epistemologies of analytic rationalism and evolutionary theory, the instrumentalism of the gun, and the accompanying ideology of human mastery over nature."[393] Later still, twentieth-century epistemologies ranging from psychoanalysis to laboratory science privileged the power of human observation, further effacing the threat formerly ascribed to the gaze of certain animals, especially predators.[394] Most recently, postmodern perspectives have rekindled some of our unease with the animal gaze. Jacques Derrida, in his probing and deeply philosophical meditation inspired by the gaze of a pet cat, calls into question our problematic withholding of subjecthood from animals, linking it to the often cruel and sadistic treatments we inflict on them.[395] Taken as a whole, Armstrong concludes, this progression has set the stage "for a new set of theoretical and material relations... [that] will allow us to learn from animals as well as about them; to encounter them with greater respect; to know when to catch their eye and when to avert our own, when to gaze intensely in the spirit of reconciliation, and when to glance away in shame."[396]

This book has focused on perceptions of wolves in mid-to-late nineteenth-century Russia during an era when modern scientific approaches were gaining ascendancy but before the genesis of contemporary ecology and our concomitant recognition of the crucial importance of apex predators in well-functioning ecosystems. As someone who supports ongoing wolf reintroduction and integration in the greater Yellowstone ecosystem and elsewhere, I have found it deeply unsettling to come to terms with how Russians of the nineteenth-century perceived their wolf populations, and also to realize that the few other Westerners who have examined this topic have utilized it to promote a vehemently negative message about wolves in the twenty-first century.[397] Looking at Yellowstone's Druid pack through a telephoto lens a few summers ago, I also had to recognize—however—that the way in which we can choose to direct our gaze on wolves like these from afar bears little resemblance to the direct and often violent interactions between wolves and humans that took place in nineteenth-century Russia. Unlike the North American experience historically and today, wolves in imperial Russia did regularly kill people, leaving an enduring imprint on Russian culture to the present day.[398]

Armstrong's survey of human interpretations of the animal gaze also underscores a crucial truth: that the looks we exchange with other species encapsulate fundamental aspects of our own identities and our understandings of the beings that surround us. While I have foregrounded moments in which Russians gazed into the eyes of wolves and saw either a savage "other" or, less commonly, a reflection or refraction of themselves, the "savage gaze" of this book's title should also be read quite pointedly. In nineteenth-century Russia, the way in which humans interpreted the gaze of this feared predator became a crucial symbol and marker of what it meant to be Russian. The manners in which Russians viewed, represented, abused, and pursued wolves were interwoven with their developing senses of who they were as Russians, the fault lines that existed in imperial society, and an overriding sense of "otherness" in relation to Western Europe. In other words, the ways in which Russians related to and described the significance of wolves during successive eras of the imperial period as Russian nature, society and culture itself evolved corresponded to deep truths (and misgivings) about themselves. Experts like Turkin and Sabaneev, hunters, advocates for animal welfare and many others interwove their perceptions of Russia's Wolf Problem with their concerns about the empire's fate in the wake of emancipation, for example, to a degree that came as a surprise to me as I became more and more familiar with the historical materials at hand. The sense of inferiority and backwardness—of shame for their country—that so clearly showed in Sabaneev's zoological monograph on wolves and Turkin's magisterial works on Russian hunting law serve as powerful testaments to the processes through which animals can become an integral part of cultural self-definition, interpretation, and representation. It is no surprise that writers like Tolstoy, Chekhov, Zinovieva-Annibal and the others I have discussed utilized wolves as central elements of their literary narratives, symbolic foci for expressing fundamental truths about the Russian empire and its people.

While the Wolf Problem came to serve as a symbol of Russian backwardness, social disorder, cultural deficiencies and inadequate government support for its citizenry, imperial Russia's wolves also represented a source of national identity and even pride. The prowess of the archetypal Russian borzoi hunter, able to bind a mature wolf single-handedly, would not have existed without the continued presence of this fearful foe. After emancipation Russian gun hunters deeply valued the experience of hunting wolves. The borzoi wolfhounds themselves, so inextricably interwoven in their characteristics and even names with their ferocious quarry, became a central symbol of Russian culture for those who admired them.

Yet here again, this source of pride for some served as an emblem of Russian cruelty for others. While we have encountered many depictions of the agony that borzois could inflict on wolves, one of the best-known passages involving Russian borzois occurs in a novel that has little to do with either wolves or hunting. Most readers of F. M. Dostoevsky will remember Ivan Karamazov's narration to his idealistic younger brother Alyosha of a story that he presents as "typical" of old Russia at the beginning of the nineteenth century. Ivan's story, which he claims to have encountered in a contemporary journal that presents archival documents, is about a general who owns 2,000 serfs and hundreds of hunting dogs in the early 1800s. One day a child house-serf throws a rock and hurts the paw of a favored hound. In response, the general has the eight-year-old boy stripped naked and unleashes his entire pack of wolfhounds on him. They tear him to pieces in front of his mother and the other serfs. After telling Alyosha the story, Ivan asks if the general deserved to be shot, to which Alyosha responds that he did.[399] Dostoevsky chose to utilize a story about Russian wolfhounds to encapsulate a central moral dilemma in his earthy yet metaphysical novel about the fundamental nature of Russia and its inhabitants.[400] Worlds apart from Tolstoy's portrayal of Nikolai Rostov's wolf hunt in *War and Peace* but set in the same era, the parable voiced by Dostoevsky's brooding intellectual Ivan demonstrates the deep significance of both wolves and their domesticated counterparts in Russian culture as a locus for self-understanding.

This story from *Brothers Karamazov* also raises another set of issues that I have encountered in writing this book. Like animals, Russia's serfs and even emancipated peasants rarely produced documents or other forms of self-expression and self-definition that have stood the test of time and could be woven into my own narrative.[401] We have encountered their perspectives mostly in indirect and mediated form. Tolstoy's Danilo, the "untrustworthy" peasants who would lie about the locations of wolves to huntsmen or naïvely fall victim to their "stratagems," the "simple folk" who would put urine and salt on their wolf bites in hopes of avoiding rabies, the Smolensk peasants who traveled to Pasteur for treatment and created a sensation, Kuzminskaya's uneducated widow—our perceptions of all of these figures rely on the interpretations of the mostly gentry and professional writers on whom I've necessarily had to draw in crafting my own narrative. Dostoevsky's parable of a sadistic serf owner who sets his borzois on a peasant boy, which is diametrically opposed to the archetype exemplified in Boriatinsky's patriarchal protection of his serfs from a wolf, shows how heavily the interpreter can skew our understanding of the relationships that truly existed.

These challenges and complications, ironically, are what have made this study so deeply satisfying for me. Russia's wolves were inextricably interwoven with imperial Russians' views of themselves and their country's place in the world. It stands to reason that their interpretations of them were creative and sometimes involved distortions, just as their interpretations of themselves and their own culture did. The ways in which they incorporated wolves into the larger project of comparing their empire to Western Europe made this all the more interesting. Wolves also remained a vital presence in imperial Russia to its very end, long after they had disappeared or dwindled throughout most of Western Europe. When three of Tolstoy's sons posed in what appears to be a studio shot with the carcasses of nine wolves they had shot during a gun hunt near Kaluga in the winter of 1900 to 1901, they were demonstrating that wild nature remained a vital part of their countryside but also that they had dominated it by killing its ultimate incarnation—a large pack of wolves (and, incidentally, that their father had not successfully passed on his renunciation of violence against animals to at least some of his progeny). Ultimately, I hope that this book has demonstrated how significant a role such an extraordinary animal can play in a culture's sense of itself.

Figure 14 Ilya, Mikhail, and Andrei Tolstoy with companions after a wolf hunt (c. 1901). Chronicle/Alamy Stock Photo

Endnotes

NOTES FOR INTRODUCTION

1 "Strashnyi sluchai," *Zhurnal okhoty* 8, no. 44 (1862): 102–4.
2 Ibid., 102–3.
3 Commoners, including emancipated peasants, were not forbidden from owning firearms, but their cost as well as that of ammunition often proved prohibitive and they were not typically at hand during a wolf attack.
4 I am following the lead of scholars like Jane Costlow and Ekaterina Pravilova in their translations of the parallel phrase *lesnoi vopros* as the "Forest Question" with capital letters but have chosen to convey the strongly negative tenor of discussions about wolves in this era by translating *vopros* as problem, which is also one of its accepted meanings. See Jane Costlow, "Geographies of Loss: The Forest Question in Nineteenth-Century Russia," in *Heart-Pine Russia Walking and Writing the Nineteenth-Century Forest* (Ithaca, NY: Cornell University Press, 2013), 81–116, and Ekaterina Pravilova, "Forests, Minerals, and the Controversy over Property in Post-Emancipation Russia," in *A Public Empire: Property and the Quest for the Common Good in Imperial Russia* (Princeton, NJ: Princeton University Press, 2014), 55–92.
5 The article and accompanying preface by Tolstoy appeared without Chertkov's full name as V. Ch-v, "Zlaya zabava," *Novoe vremia* 5284 (November 13, 1890): 2–3.
6 Ecocriticism has historical roots but has become a recognized school of literary criticism only in recent decades, especially with the 1993 establishment of the peer-reviewed journal *Interdisciplinary Studies in Literature and the Environment (ISLE)* and its sponsoring society, the Association for Study of Literature and Environment (ASLE). For some foundational texts and overviews of the field see C. Glotfelty and H. Fromm, eds., *The Ecocriticism Reader: Landmarks in Literary Ecology* (Athens: University of Georgia Press, 1996); G. Carr, ed., *New Essays in Ecofeminist Literary Criticism* (Lewisburg, PA: Bucknell University Press, 2000); S. Rosendale, *The Greening of Literary Scholarship: Literature, Theory, and the Environment* (Iowa City: University of Iowa Press, 2002); Greg Garrard, *Ecocriticism* (London: Routledge, 2004); and Timothy Clark, *The Cambridge Introduction to Literature and the Environment* (Cambridge: Cambridge University Press, 2011).
7 Animal studies has one set of roots in the literature of animal welfare and animal rights pioneered by such authors as Peter Singer in his *Animal Liberation: A New Ethics for Our Treatment of Animals* (New York: Random House, 1975) but has recently gained momentum and broadened in various directions. Activist scholarship, which asserts that description and analysis are not in themselves adequate responses to the injustices that humanity has perpetrated on animals, continues to represent a vital component of animal studies, for which many of its proponents prefer the term "critical animal studies."

A prime example is the *Journal for Critical Animal Studies*, which "was designed to develop activist consciousness of animal liberation history, practice, theory, and politics, while also encouraging Critical Animal Studies (CAS) scholarship" (see http://journalforcritical-animalstudies.org/). Another serial publication worthy of note is the *Society & Animals Journal*, established—like *ISLE*—in 1993, and the companion Human–Animal Studies Book Series, also published by the Animals & Society Institute in Ann Arbor, Michigan, (see https://www.animalsandsociety.org/human-animal-studies/). The field of human–animal studies is expanding rapidly on multiple fronts as its practitioners draw on their own varied academic backgrounds and levels of comfort in balancing scholarly "objectivity" and moral fervor, as well as the challenges of working through and across disciplines. For an incisive recent overview of the field that highlights its "meta-disciplinary" nature see Garry Marvin and Susan McHugh, "In It Together: An Introduction to Human–Animal Studies," in *Routledge Handbook of Human–Animal Studies*, eds. Garry Marvin and Susan McHugh (London: Routledge, 2014), 1–9. A small selection of other helpful sources includes Nigel Rothfels, ed., *Representing Animals* (Bloomington: Indiana University Press, 2002); Aaron Gross and Anne Vallely, eds., *Animals and the Human Imagination: A Companion to Animal Studies* (New York: Columbia, 2012); and Nik Taylor and Richard Twine, eds., *The Rise of Critical Animal Studies: From the Margins to the Centre* (London: Routledge, 2014). I will refer to many more individual works during the course of the book.
8   Jane Costlow and Amy Nelson, eds., *Other Animals: Beyond the Human in Russian Culture and History* (Pittsburgh, PA: University of Pittsburgh Press, 2010), 4. This volume represents the most ambitious attempt to address the significance of animal studies for interpreting Russian culture and to gather a richly interdisciplinary set of perspectives on the subject and includes my chapter as Ian M. Helfant, "That Savage Gaze: The Contested Portrayal of Wolves in Nineteenth-Century Russia," 63–76. Another recent work of interest, which features some papers by Russian scholars alongside European peers, is Léonid Heller and Anastasiia Vinogradova de La Fortelle, eds., *Zveri i ikh reprezentatsii v russkoi kul'ture: Trudy Lozannskogo simpoziuma, 2007 g.* (St. Petersburg: Baltiiskie sezony, 2010).
9   See note 4.
10 See Henrietta Mondry, *Political Animals: Representing Dogs in Modern Russian Culture* (Leiden, The Netherlands: Brill Rodopi, 2015).
11 A small selection of other ecocritically-oriented works that focus on imperial Russia include C. D. Ely, *This Meager Nature: Landscape and National Identity in Imperial Russia* (DeKalb: Northern Illinois University Press, 2003); Thomas Newlin, "At the Bottom of the River: Forms of Ecological Consciousness in Mid-Nineteenth-Century Russian Literature," *Russian Studies in Literature* 39, no. 2 (2003): 71–90; Arja Rosenholm and Sari Autio-Sarasmo, eds., *Understanding Russian Nature: Representations, Values and Concepts* (Helsinki: University of Helsinki Aleksanteri Institute, 2005); Ian M. Helfant, "S. T. Aksakov: The Ambivalent Proto-Ecological Consciousness of a Nineteenth-Century Russian Hunter," *Interdisciplinary Studies in Literature and Environment* 13, no. 2 (Summer 2006): 57–71; and Thomas Newlin, "Swarm Life" and the Biology of *War and Peace*," *Slavic Review* 71, no. 2 (2012): 359–84. I will refer to additional works later.
12 See John D. C. Linnell et al., eds., *The Fear of Wolves: A Review of Wolf Attacks on Humans* (Trondheim, Norway: NINA-NIKU, 2002), 3. This rigorously researched multi-authored work analyzes the history of wolf attacks primarily from the eighteenth century through the present. It attempts to provide a world-wide overview and assesses Russia's current wolf population (~40,000) as the world's largest, outnumbered only by the total in all of North America (~60,000), 18, 24. It ascribes the vast majority of historical wolf attacks worldwide

to rabid wolves, emphasizing the particular rarity of aggression by non-rabid wolves toward humans in North America while noting that such attacks appear to have been more prevalent in Russia, 24–25, 28–31.

13 See L. David Mech and Luigi Boitani, eds., *Wolves: Behavior, Ecology, and Conservation* (Chicago: University of Chicago Press, 2003). The volume does touch in a few places on Russia's wolves primarily through reference to works by the Soviet scholar D. I. Bibikov, to whom I shall return shortly.

14 See Garry Marvin, *Wolf* (London: Reaktion Books 2012), 131, n. 14.

15 See Patrick Masius and Jana Sprenger, eds., *A Fairytale in Question: Historical Interactions between Humans and Wolves* (Cambridge, UK: The White Horse Press, 2015). I will cite individual contributions to this volume later.

16 D. I. Bibikov, ed., *Volk: Proiskhozhdenie, sistematika, morfologiia, ekologiia* (Moscow: Nauka, 1985). Bibikov also edited the proceedings of a 1987 conference on wolves held in Moscow, which focused especially on means of controlling wolf populations: D. I. Bibikov, ed., *Ekologiia, povedenie i upravlenie populiatsiiami volka. Sbornik nauchnykh trudov* (Moscow: VASKhNIL, 1989).

17 The translation appeared as D. I. Bibikov, ed., *Der Wolf: Canis Lupus*, trans. Günther Grempe (Wittenberg Lutherstadt, Germany: A. Ziemsen, 1988).

18 M.P. Pavlov, *Volk*, 2nd ed. (Moscow: Agropromizdat, 1990).

19 See Pavlov, "Opasnost' volka dlia liudei," in *Volk*, 136–69. For an efficient critique of Pavlov's methodology and results see Linnell, *Fear of Wolves*, 24–25. I shall be dealing with extensive primary sources relating to wolf attacks in the tsarist period during the course of this book, so will not engage directly in the debate here.

20 Will N. Graves' *Wolves in Russia: Anxiety Through the Ages* (Calgary, Canada: Detselig, 2007). Graves argues that Russia's large wolf populations demonstrate how destructive wolves can be when insufficiently controlled. He also contends that wolves carry numerous parasites and diseases that pose threats to other species, and that even non-rabid wolves regularly attack humans without provocation. These appraisals diverge from those of the overwhelming majority of contemporary wildlife biologists. For an evaluation of the 2007 book by a wildlife biologist see James H. Shaw, "Review of *Wolves in Russia: Anxiety through the Ages*, by Will N. Graves," *The Journal of Wildlife Management* 73, no. 6 (2009): 1025–26. Graves, who is not an academic but rather learned Russian as a linguist for the US Air Force and developed an abiding interest in Russia's wolves, has also recently co-written a book on wolves with trial lawyer Ted Lyon, self-published by the latter. See Ted B. Lyon and Will N. Graves, *The Real Wolf: The Science, Politics, and Economics of Coexisting with Wolves in Modern Times* (Billings, MT: Ted B. Lyon [Farcountry Press], 2014). The book argues against ongoing efforts to reintroduce and extend protections for wolves in parts of the continental United States. It includes two brief chapters by Graves on wolves in Russia which, like his earlier book, cite sources—even for direct quotes and figures—only sporadically and contain patently false statements such as: "It is well known that victims bitten by animals carrying rabies face a 100% chance of dying unless they receive treatment before symptoms appear" (Lyon, *Real Wolf* [Kindle edition], loc. 1613). In reality, the chances of developing rabies after being bitten by a rabid animal vary considerably depending on the nature of the wounds and other factors, as I shall discuss more thoroughly in my chapter on rabies. Given the flawed and biased nature of Graves work, I shall refer to it only sparingly in the course of my study. For a review of the 2014 book that notes some of its shortcomings, see Adrian P. Wydeven, "Review of *The Real Wolf: The Science, Politics, and Economics of Coexisting with Wolves in Modern Times*," *The Journal of Wildlife Management* 80, no. 7 (2016): 1334–35. Wydeven, who served from 1990 through 2013 as the

head of Wisconsin's wolf recovery and management program, concludes: "Because of the lack of science and misleading information, I would have a hard time recommending this book. In the end, *The Real Wolf* is mostly a political book that attempts to create negative perspectives on wolves and gain support for such negative perspectives," 1335.

21 See *Wolk 1 Der Lasarewski-Report Zur Wolf In Rußland*, comp. W. Rathgeber and V. Bonvicini (Munich: Bengelmann Verlag and W. Rathgeber, 2011). The book provides a valuable service in making Lazarevsky's 1876 text, which is a bibliographic rarity that I originally encountered while doing primary research in Russia, readily available. Like Graves' work, it aims to undermine the belief that wolves and humans can coexist even in modern times. For example, one of its principal authors, German physician and retired academic Walter Rathgeber, expresses vehement opposition to any resurgence of wolves in Europe, arguing that "the reintroduction of wolves into our cultural landscape is tantamount to the breeding of locusts protected by the law," 165. He concludes: "The Wolf's survival is sponsored by **our assets**, our **wealth, our overtime, and our taxes**. Yet, the real price is the loss of our **social norms** and **values**, and, last but not least, our **public and individual health. In other words, the price is high and ultimately a sacrifice of 'our way to happiness,'**" 169 [boldface in original]. I will discuss Lazarevsky's brochure in detail in Chapter 2.

22 Jon T. Coleman, *Vicious: Wolves and Men in America* (New Haven, CT: Yale University Press, 2004).

23 Brett L. Walker, *The Lost Wolves of Japan* (Seattle: University of Washington Press, 2005). Both of these works, as well as my own, owe a debt to Barry Holstun Lopez, *Of Wolves and Men* (New York: Charles Scribner's Sons, 1978), which provides a now classic account of human-wolf relations throughout history and across civilizations. Lopez's sensitive portrayal of wolves devotes particular attention to the ways in which the indigenous people of the Alaskan Arctic coexisted with wolves over the centuries, a perspective that diverged dramatically from the dominant tradition of negative depictions of wolves and helped to set the stage for the ongoing contemporary reappraisal of wolves. See also David Quammen, *Monster of God: The Man-Eating Predator in the Jungles of History and the Mind* (New York: W. W. Norton, 2003), for a roving and thoughtful consideration of the ways in which human societies have construed their relationships with large predators.

24 See Ian M. Helfant, *The High Stakes of Identity: Gambling in the Life and Literature of Nineteenth-Century Russia* (Evanston, IL: Northwestern University Press, 2002).

25 Erica Fudge, "A Left-Handed Blow: Writing the History of Animals," in Rothfels, *Representing Animals*, 5.

26 Ibid., 6.

27 Kay Peggs explores this tendency within Critical Animal Studies in "From centre to margins and back again: critical animal studies and the reflexive human self," in Taylor, *The Rise of Critical Animal Studies*, 36-51. Drawing on the notion of "intersectionality" and approaches taken by earlier feminist scholars, Peggs notes that "CAS writers are not indistinct figures; rather, through autobiographical disclosures, they move from the margins to the center of their work," 46.

28 Fudge, "A Left-Handed Blow," 8.

29 Ibid., 8–11, presents her overview of the three categories and additional scholarly works that exemplify them.

30 John Berger, "Why Look at Animals?" in *About Looking* (New York: Pantheon, 1980), 5.

31 Philip Armstrong, "The Gaze of Animals," in *Theorizing Animals: Re-Thinking Humanimal Relations*, eds. Nik Taylor and Tania Signal (Leiden, The Netherlands: Brill 2011), 175–99.

NOTES FOR CHAPTER 1

32 P. M. Machevarianov, *Zapiski psovogo okhotnika Simbirskoi gubernii*, ed. A. V. Skomorokhova (Minsk: Polifakt, 1991), 11.
33 L. N. Tolstoy, *Polnoe sobranie sochinenii v 90 tomakh* (Moscow: Gosudarstvennoe izdatel'stvo khudozhestvennoi literatury, 1928–64), 10: 254; L. N. Tolstoy, *War and Peace*, trans. Richard Pevear and Larisa Volokhonsky (New York: Alfred A. Knopf, 2007), 501. All subsequent citations will provide page numbers for the Russian text first, followed by the Pevear and Volokhonsky translation. Gary Saul Morson focuses on this "happiest moment" of Nikolai's life as a quintessential example of Tolstoy's authorial certitude in his *Hidden in Plain View: Narrative and Creative Potentials in 'War and Peace'* (Stanford, CA: Stanford University Press, 1987), 156–57, as does Andrew D. Kaufman in his *Give War and Peace a Chance: Tolstoyan Wisdom for Troubled Times* (New York: Simon & Schuster, 2014), 106. I should point out that Aylmer Maude translates this and subsequent passages utilizing female pronouns for the wolf (Tolstoy, *War and Peace*, ed. George Gibian [New York: Norton, 1996], 442–44). In the original Russian the wolf's gender is arguably indeterminate, as it could be either the dominant male or female of the pack, but I will follow Pevear and Volokhonsky in utilizing masculine forms. The Russian *volk* (wolf) is a masculine noun but is often used for wolves of unspecified gender, although a female equivalent—*volchitsa*—can be used to emphasize that a particular animal is female and likely would be at some point during these pages by Tolstoy if that were the case. Russians typically use masculine or feminine pronouns and possessive adjectives corresponding to "he/she" and "his/her" rather than the neuter equivalents of "it/its" for animals other than those designated through neuter identifiers such as "insect" (*nasekomoe*). For an intriguing linguistic analysis of gender patterns in the names of Russian animal species that categorizes animals in part by the danger they pose to humans, see Alexander V. Kravchenko, "The Cognitive Roots of Gender in Russian," *Glossos* 3 (Spring 2002): 1–13.
34 See Linnell, *Fear of Wolves*, 18, for a chart of wolf populations across Europe, North America, and Russia from the eighteenth century to c. 2000. Wolves were exterminated in the 1680s in England, in the 1770s in Ireland and Denmark, and experienced dramatic reductions of population or eradication in other European countries over the course of the nineteenth and twentieth centuries. I will return to this topic in Chapter 2.
35 M. Volunin, "Zver', borzoi volkodav prinadlezhashchy Brigadiru Kniaziu Gavrilu Fedorovichu Boriatinskomu," *Zhurnal konnozavodstva i okhoty* 1, no. 2 (1842): 63–79.
36 Ibid., 73.
37 Marvin and McHugh propose "the interrelated and contested concepts of wild, domesticated, and feral, which suggest a crude, yet useful, set of coordinates" as an ordering structure for conceptualizing the range of works in animal studies presented in their handbook ("In It Together: An Introduction to Human–Animal Studies," 3). The borzoi's liminality makes it a fascinating instrument for probing these problematic and porous borders, as do accounts of attempts to domesticate wolves that I will address later.
38 Tolstoy, *PSS*, 10: 259; Tolstoy, *War and Peace*, 506.
39 Tolstoy, *PSS*, 10: 258; Tolstoy, *War and Peace*, 505.
40 See Garry Marvin, "Unspeakability, Inedibility, and the Structures of Pursuit in the English Foxhunt," in *Representing Animals*, ed. Nigel Rothfels (Bloomington: Indiana University Press, 2002), 139–58, for an incisive "thick description" of English fox hunting. See also his "Cultured Killers: Creating and Representing Foxhounds," *Society & Animals* 9, no. 3 (2001): 273–92 and his "A Passionate Pursuit: Foxhunting as Performance," *Sociological*

Review 51 (2003): 46–60. Marvin focuses partly on English foxhunting in the modern era but provides insight into foxhounds as bearers of cultural meaning that I have found useful for thinking about the significance of borzois in Russian culture. For detailed portrayals of foxhunting in nineteenth-century England see David C. Itzkowitz, *Peculiar Privilege: A Social History of English Foxhunting*, 1753-1885 (Hassocks, UK: Harvester Press, 1977) and Raymond Carr, *English Fox Hunting: A History*, rev. ed. (London: Weidenfeld and Nicolson, 1986).

41 Christopher Ely explores the reactions of Western travelers, as well as Russians, to the vast distances and wild landscapes of the Russian empire in *This Meager Nature* (see note 11).

42 Craig Scott differentiates the "hunt for the wolf, portrayed as a natural affirmation of life with a truth and grandeur all its own," from the "trifling competitiveness" of the later hunt for a fox and "rabbit" in his "The Hunt for Truth in War and Peace," *Tolstoy Studies Journal* 3 (1990): 121. Unfortunately, Scott's brief treatment is limited by his apparent ignorance of the hunt as a social and historical institution in early nineteenth-century Russia, his lack of basic knowledge about the ways in which borzois were utilized, and by specific lapses. For example, he misidentifies the large and swift European hare (*zayats*) that the competing borzois pursue, which was considered respectable quarry, as a smaller and much less challenging rabbit (*krolik*), 122. Kaufman focuses on the link between Nikolai's happiness, transcendence, and faith as he experiences the hunt in *Give War and Peace a Chance*, 103–11.

43 The work was republished in 1930 and again in 1985. I have drawn on the scholarly introductions to both of these editions for details concerning Driansky's literary career and what little is known of his biography. See P. E. Shchegolev, "Ob avtore *Zapisok Melkotravchatogo*" in Ye. E. Driansky, *Zapiski melkotravchatogo*, ed. P. E. Shchegolev (Moscow: ZIF Zemlia i fabrika, 1930), 3–35 and V. M. Guminsky, "Predislovie" in Ye. E. Driansky, *Zapiski melkotravchatogo*, ed. V. M. Kurganova (Moscow: Sovetskaya Rossiia, 1985), 3–20. The work has never been translated. My translations will cite the 1985 edition.

44 Guminsky points this out as well in his "Predislovie," 14.

45 N. M. Reutt, *Psovaya okhota*, 2 vols. (St. Petersburg: Tipografiia Karla Kraiia, 1846). A contemporary reviewer of the work extolled it as the first significant treatment of hunting with scent hounds and borzois in Russian. The anonymous reviewer emphasized Reutt's familiarity with both classical works and contemporary Western European treatments of hunting, as well as his enormous practical experience. See "*Psovaya okhota. Sochinenie N. Reutta* [review]," *Biblioteka dlia chteniia* 76, no. 6 (1846): 24–27.

46 Reutt, *Psovaya okhota*, 1: 25.

47 The work was republished in 1991 (see note 32). See P. Semchenkov, "Posledny iz mogikan [Posleslovie]," in Machevarianov, *Zapiski*, 143–52, for a helpful overview of Machevarianov's significance that I have drawn on in my own summary.

48 Machevarianov, *Zapiski*, 11. Machevarianov wrote the foreword in 1860 (the year after the publication of Driansky's complete *Notes*).

49 P. M. Gubin, *Polnoe rukovodstvo ko psovoi okhote v trekh chastiakh*, 3 vols. (Moscow: Tipo-litografiia Morits Ivanovich Neiburger, 1890).

50 Gubin, "Predislovie," *Polnoe rukovodstvo*, iii. The three volumes of Gubin's work total over 450 pages. For his ten chapters devoted specifically to hunting wolves with borzois see *Polnoe rukovodstvo*, 3: 58–102.

51 Gubin summarizes some of these concerns in his "Zakliuchenie" (Conclusion), *Polnoe rukovodstvo*, 3: 148–52.

52 See Helfant, "S. T. Aksakov," for a discussion of the ways in which Aksakov interwove personal experience, concern for Russia's dwindling wildlife, and ornithological observations

into a memoiristic subgenre of nature writing. I will not delve into the impassioned debate among these writers and others of their generation about the relative merits of hunting with packs of hounds and borzois versus bird-hunting with a single dog (the specialty of Aksakov and his contemporary Ivan Turgenev). One of the most sensitive current interpreters of that form of hunting in the Russian context is Thomas Hodge. See, for example, his "Ivan Turgenev on the Nature of Hunting," in *Words, Music, History: A Festschrift for Caryl Emerson, Part One*, ed. L. Fleishman, G. Safran, and M. Wachtel (Stanford, CA: Stanford University Press, 2005), 291–311. I have chosen not to include L. P. Sabaneev's substantial writings on borzois and other hunting dogs as one of this chapter's central sources partly for tactical reasons, as much of Chapter 2 will be devoted to his uniquely important monograph on wolves.

53 For a fuller discussion of the gambling scene see Helfant, *The High Stakes of Identity*, 120–21.
54 Tolstoy, *PSS*, 10: 57; Tolstoy, *War and Peace*, 340.
55 Donna Tussing Orwin compares Dolokhov to the partisan Tikhon and Tikhon to a wolf in his ferocity, then observes that this links Tikhon with the wolf that Rostov hunts. I agree with her but think the direct parallel between Dolokhov and the wolf is equally productive, given his aggression toward Nikolai and Nikolai's conviction that he can retrieve the honor Dolokhov has imperiled by vanquishing the wolf. See Orwin, See Orwin's *Tolstoy's Art and Thought, 1847-1880* (Princeton, NJ: Princeton University Press, 2013). ProQuest ebrary. Web. August 5, 2017, 114.
56 Along these lines and in keeping with Marvin's and McHugh's considerations about the domestic, feral, and wild Guminsky observes in his foreword to Driansky's *Notes*: "Here, between the human and the beast there stands, in essence, yet another beast, yet one that is to a greater or lesser degree tamed, domesticated and who therefore supports the human side. The basic struggle unfolds between the representatives of one or almost one world, while the human is primarily an interested observer of the borzoi hunt, and then afterwards a participant in its finale" (Guminsky, "Predislovie," 9).
57 For further discussion of Tolstoy's turn to vegetarianism see Ronald D. LeBlanc, "Tolstoy's Way of No Flesh: Abstinence, Vegetarianism, and Christian Physiology," in *Food in Russian History and Culture*, ed. Musya Glants and Joyce Stetson Toomre (Bloomington: Indiana University Press, 1997), 81–102.
58 Tolstoy, *PSS*, 60: 302.
59 Tolstoy, *PSS*, 60: 306.
60 Tolstoy, *PSS*, 60: 314.
61 Tolstoy, *PSS*, 48: 427.
62 Tolstoy, *PSS*, 10: 244; Tolstoy, *War and Peace*, 493.
63 Machevarianov, *Zapiski*, 18.
64 Tolstoy, *PSS*, 10: 244; Tolstoy, *War and Peace*, 493.
65 Tolstoy, *PSS*, 10: 245; Tolstoy, *War and Peace*, 493–94.
66 For a thorough analysis of the significance of Cossacks in Russian literature and culture see Judith Deutsch Kornblatt, *The Cossack Hero in Russian Literature: A Study in Cultural Mythology* (Madison: University of Wisconsin Press, 1992).
67 Each of the three hunting guides elucidates the roles of the various hunters in larger hunting establishments, which could overlap or be combined in lesser ones. These included the *lovchy* (head huntsman) who had overall responsibility; the *doezzhachy* (head kennelman), whose primary responsibility was to train and oversee the scent hounds, sometimes with the help of assistants called *psari* (assistant kennelmen) and *vyzhliatniki* (whippers-in); the *stremiannyi* (borzoi handlers), who had oversight over the master's principal borzois;

and the *borziatniki* (assistant borzoi handlers), who were in charge of additional borzois. Machevarianov provides the most coherent overview of their different but sometimes overlapping responsibilities in his *Zapiski*, 19–22. See also Gubin's complementary explanation of the various hunters' titles and roles in his *Polnoe rukovodstvo*, 1: 10–16. Danilo is identified as the Rostovs' "doezzhachy i lovchy," which Pevear and Volokhonsky translate accurately as "head kennelman and huntsman." This indicates that he has overall responsibility both for the hounds and the conduct of the hunt itself.

68 Machevarianov, *Zapiski*, 21–22.
69 Ibid., 19.
70 See Gubin, *Polnoe rukovodstvo*, 2: 138–48 for a full range of examples which include the repeated staccato cry of *U-liu-liu-liu!* that hunters used to encourage borzois unleashed on a wolf.
71 Driansky, *Zapiski*, 84.
72 Gubin, *Polnoe rukovodstvo*, 2: 63.
73 Gubin, *Polnoe rukovodstvo*, 3: 150.
74 For an overview of gentry landholdings in the wake of the emancipation and the decades that followed see Terence Emmons, "The Russian Landed Gentry and Politics," *Russian Review* 33, no. 3 (July 1974): 269–83.
75 Tolstoy, *PSS*, 10: 245; Tolstoy, *War and Peace*, 494.
76 Tolstoy, *PSS*, 10: 245–46; Tolstoy, *War and Peace*, 494.
77 Tolstoy, *PSS*, 10: 246; Tolstoy, *War and Peace*, 494.
78 Orlando Figes begins his *Natasha's Dance: A Cultural History of Russia* (New York: Picador, 2002) by focusing on Natasha's graceful folk dance when she and Nikolai visit "Uncle" at his simple home after the hunt, xxi–xxviii. He argues that this embodies Tolstoy's yearning for a genuinely Russian community that transcends class, which Natasha comes to symbolize.
79 See N. V. Turkin, *Okhota i okhotnich'e zakonodatel'stvo v 300-letny period tsarstvovaniia doma Romanovykh* (Moscow: Imperatorskoe Obshchestvo razmnozheniia okhotnich'ikh i promyslovykh zhivotnykh i pravil'noi okhoty, 1913), for a description of the royal hunting establishment during the successive reigns of the Romanov dynasty. The female empresses who ruled for most of the eighteenth century were especially ardent hunters, 71–85.
80 Tolstoy, *PSS*, 10: 247; Tolstoy, *War and Peace*, 495
81 Gubin, *Polnoe rukovodstvo*, 1: 8–9.
82 Reutt, *Psovaya okhota*, 2: 2.
83 Tolstoy, *PSS*, 10: 248; Tolstoy, *War and Peace*, 496. See Marvin's explanation of the disdain shown to those who refer to foxhounds as "dogs" in his "Cultured Killers," 279.
84 Tolstoy, *PSS*, 10: 248; Tolstoy, *War and Peace*, 496.
85 Gubin, *Polnoe rukovodstvo*, 1: 130–34.
86 Gubin, *Polnoe rukovodstvo*, 3: 56–57.
87 Ibid., 101.
88 Tolstoy, *PSS*, 10: 249; Tolstoy, *War and Peace*, 497.
89 Ibid.
90 Tolstoy, *PSS*, 10: 251; Tolstoy, *War and Peace*, 498–99.
91 Tolstoy, *PSS*, 10: 251; Tolstoy, *War and Peace*, 499.
92 Craig, "The Hunt for Truth," also highlights Danilo's criticism of the old count but ascribes it to the equalizing power of nature rather than their vastly different levels of hunting expertise and dedication to the outcome of the hunt, 121.
93 Reutt, *Psovaya okhota*, 2: 56–57. For a related elucidation of the "voice" of the pack in modern foxhunting see Marvin's "Cultured Killers," 282–84.

94 Tolstoy, *PSS*, 10: 252; Tolstoy, *War and Peace*, 499–500. Kaufman emphasizes the intensity of Nikolai's awareness in this scene and the texture of individual happiness in such moments of intense consciousness (*Give War and Peace a Chance*, 105–11). He touches on some of the same passages I address, but with a very different interpretive agenda.
95 Tolstoy, *PSS*, 10: 252–53; Tolstoy, *War and Peace*, 500.
96 Tolstoy, *PSS*, 10: 253; Tolstoy, *War and Peace*, 501. Pevear and Volokhonsky translate the Russian phrase as "Halloolooloo." Reutt states that borzois must be familiar with at least seven commands (including this one), and, like Gubin, differentiates between those utilized in hunting predators—namely wolves and foxes—and hares, as well as those used in both cases (Reutt, *Psovaya okhota*, 1: 118–19).
97 Tolstoy, *PSS*, 10: 253–54; Pevear and Volokhonsky, 501.
98 Machevarianov, *Zapiski*, 98.
99 Tolstoy, *PSS*, 10: 255; Tolstoy, *War and Peace*, 502.
100 Tolstoy, *PSS*, 10: 255; Tolstoy, *War and Peace*, 502.
101 Driansky, *Zapiski*, 110. The borzoi's name corresponds to the imperative of the verb *porazhat'* (to strike, rout, defeat), which was often paired with the instrumental form of the noun for dagger to mean "stab with a dagger" (*porazhat' kinzhalom*).
102 Driansky, *Zapiski*, 41.
103 Reutt, *Psovaya okhota*, 2: 187–88.
104 Ibid., 188–89
105 See especially Marvin, "Cultured Killers," 273–76.
106 Marvin, "Unspeakability," 143.
107 Tolstoy, *PSS*, 11: 63-65; Tolstoy, *War and Peace*, 652–54.
108 Tolstoy, *PSS*, 11: 66; Tolstoy, *War and Peace*, 655. For Kaufman's analysis of the scene see *Give War and Peace a Chance*, 154–56.

NOTES FOR CHAPTER 2

109 N. V. Turkin, *Zakony ob okhote. Kriticheskoe issledovanie russkikh okhotnich'ikh zakonopolozhenii* (Moscow: Redaktsiia zhurnala *Priroda i okhota*, 1889), 76.
110 Pravilova, *A Public Empire*, 57. Pravilova's deeply researched book probes the changing and contested definitions of property ranging from Russia's natural resources (including wildlife) to its historical monuments and artistic treasures in this era. I shall return to it later in the chapter. For an extensive overview of the *zemstvo* as a form of representative government that nevertheless continued to privilege the landowning gentry see Roberta Thompson Manning, *The Zemstvo in Russia: An Experiment in Local Self-Government* (Cambridge: Cambridge University Press, 1982).
111 A history of the Moscow Society can be found in Moskovskoe Obshchestvo Okhoty imeni Imperatora Aleksandra II, *Al'bom v pamiat' piatidesiatiletniago iubileia Moskovskago Obshchestva Okhoty imeni Imperatora Aleksandra II (1862-1912)*, comp. A. A. Beer et al. (Moscow: T-vo skoropechatni A.A. Levenson, 1913). This work of over 300 pages compiles a narrative history, 40 appendices ranging from the Society's charter (*ustavy*), to letters from local authorities thanking the Society for killing wolves, to membership lists and photographs, to the compiled results of hunting dog field trials. A summary of the Society's first twenty-five years also appeared in *Nature and Hunting*: A. E. Fedorov, comp., "Materialy dlia istorii Moskovskago Obshchestva Okhoty," *Priroda i okhota*, February 1888, 1–36. My overview draws on both of these sources, which overlap in many respects

but also provide complementary perspectives and information. Where it seems necessary to reference page numbers, I will cite the first source as Moskovskoe Obshchestvo, *Al'bom* and the second as Fedorov, *Materialy*.

112 See Moskovskoe Obshchestvo, *Al'bom*, 55–56, for the 1862 charter, which identifies the Society's overriding goal as promoting "proper conceptions of hunting and oversight of correct hunting practices," specifies the dues and selection process for new members, and emphasizes the Society's goal of predator reduction.

113 Ibid., 133–38. This hunt was one of the reasons that the tsar's son, Grand Duke Sergei Alexandrovich, a major benefactor, granted the Society permission to add "in honor of Alexander II" to its official title on its thirty-fifth anniversary in 1897 (Ibid., 35).

114 Ibid., 149–77.

115 Ibid., 125–32.

116 Ibid., 55–56.

117 Ibid., 45.

118 Fedorov, *Materialy*, 32. For an extensive recent history of the development of the Russian empire's railway system see Aleksei Vul'fov, *Istoriia zheleznykh dorog Rossiiskoi imperii: Vse-taki stroit'!* (Moscow: Ripol, 2016). The country's first major line between Moscow and Petersburg opened in 1851 and the decades after serfdom saw a steady expansion of railroad lines, particularly in the western provinces of European Russia.

119 Fedorov, *Materialy*, 29–30.

120 A summary of this method of hunting wolves is included in a chapter entitled "Zimniia okhoty," Moskovskoe Obshchestvo, *Al'bom*, 179–88. It's important to note that coursing wolves with mixed packs of hounds and borzois still occurred in late imperial Russia, and that both the Moscow and other hunting societies sponsored such hunts on occasion, but that they became increasingly rare.

121 Ibid., 24, and appendices 27, 28 (107–8).

122 Ibid., 39, 49.

123 Turkin provides an overview of the Imperial Society's first twenty-five years in his densely researched "Istorichesky ocherk deiatel'nosti Imperatorskago Obshchestva razmnozheniia okhotnich'ikh i promyslovykh zhivotnykh i pravil'noi okhoty za 25-letny period sushchestvovaniia, 1872-1897," *Priroda i okhota*, January 1898, 1–115. I have not found an analogous historical summary of the following two decades, although the Imperial Society published its journal, *Nature and Hunting*, through 1912.

124 Ibid., 1–2, 71. The Russian name for the former society was *Imperatorskoe Moskovskoe obshchestvo ispytatelei prirody* (MOIP).

125 Ibid., 3–4.

126 For additional background on Sabaneev that emphasizes his ability to write about hunting in a scientifically informed manner that also appealed to a broader readership see Mary Cavender, "Hunting in Imperial Russia: State Policy and Social Order in L. P. Sabaneev's Writing," *The Russian Review* 76 (July 2017): 484–501. Unfortunately, Cavender attributes an extended editorial on hunting law that appeared in *Hunting News* in 1888, which provides the focus for much of her article, to Sabaneev. In fact, it was written by Turkin and published separately in 1889 as a monograph, as I shall discuss below.

127 Turkin, "Istorichesky ocherk," 5–10. I am relying partly on Turkin's history for this summary, as well as my own perusal of the entire thirty-year run of the journal during my 2006, 2012, and 2014 research trips to Petersburg.

128 Ibid., 13–25.

129 Ibid., 11.
130 Ibid., 38–43.
131 Ibid., 12–13.
132 For an overview of Lazarevsky's intertwined bureaucratic and literary career see G. V. Krasnov, "Lazarevsky, Vasily Matveevich," in *Russkie pisateli 1800-1917. Biografichesky slovar'*, ed. P. A. Nikolaev (Moscow: FIANIT, 1994), 3: 282–83.
133 Lazarevsky, *Ob istreblenii volkom domashniago skota i dichi i ob istreblenii volka. Prilozhenie k Pravitel'stvennomu Vestniku* (St. Petersburg: Tipografiia Ministerstva vnutrennikh del, 1876). As I mentioned in the introduction, the work has recently been republished in a German compilation that includes the full Russian text, which was previously a bibliographic rarity. See Rathgeber, *Wolk 1. Der Lasarewski-Report Zur Wolf In Rußland*, 59–137 and note 21. Because of the greater accessibility of the recent publication, I will cite it rather than the 1876 original.
134 The preceding figures can be found in the Rathgeber edition, 69–77.
135 Ibid, 90–91.
136 Ibid., 92.
137 Ibid., 117–22.
138 Ibid., 105–109.
139 Ibid., 109.
140 Ibid., 127–37.
141 Ibid., 80.
142 L. P. Sabaneev, "Volchy vopros," *Zhurnal okhoty* 5 (1876): 42–50.
143 Ibid., 47–48.
144 Ibid., 48–49.
145 Vlad Belov, "K volch'emu voprosu," *Priroda i okhota*, January 1878, 49.
146 Ibid., 50–51.
147 S. Bezobrazov, "Ocherki lesnoi okhoty v Iamburgskom uezde," *Priroda i okhota*, April 1878, 23–24.
148 O. M. "Zametka k volch'emu voprosu," *Priroda i okhota*, January 1880, 104.
149 Ibid., 104–5.
150 N. V. Mazharov, "Po povodu pechal'nykh okhot v osen' 1879 goda," *Priroda i okhota*, January 1880, 138.
151 For example, Sabaneev published "Methods of Eradicating Wolves," which totaled more than 100 pages that were later incorporated into the larger work, in successive issues of the July, August, and September 1878 issues of *Nature and Hunting*. The monograph has recently been republished as part of a collection of Sabaneev's work as L. P. Sabaneev, *Volk. Okhotnich'ia monografiia* in his *Vse ob okhote*, ed. D. Galkina (Moscow, 2011), 1340–1547. For the reader's convenience, I will cite the 2011 text rather than the original publication of 1880 or any of the earlier versions. Sabaneev also wrote extensively about both scent hounds and borzois in texts that could have featured in Chapter 1, but I chose to focus on other authorities for the sake of balance and variety. His writings on hunting dogs are also included in the 2011 edition.
152 See V. S. Shishkin, "Zarozhdenie, razvitie i preemstvennost' akademicheskoi zoologii v Rossii," *Zoologichesky zhurnal* 78, no. 12 (1999): 1381-95.
153 Sabaneev, *Volk*, 1343-44.
154 Ibid., 1345.
155 Ibid., 1347–48.

156 Ibid., 1349.
157 Ibid., 1349–50. For a broad-ranging set of essays on the figure of the vampire in Slavic and European folk beliefs, with occasional reference to werewolves, see Jan Louis Perkowski, *Vampire Lore: From the Writings of Jan Louis Perkowski* (Bloomington, IN: Slavica, 2006).
158 See notes 4 and 110.
159 See Cavender, 489–93.
160 Sabaneev, *Volk*, 1360.
161 Ibid., 1360–62.
162 Ibid., 1382–83.
163 Ibid., 1385.
164 Ibid., 1382.
165 Ibid., 1386–87.
166 Ibid. Later in his monograph Sabaneev explicitly accused gentry borzoi hunters of perpetuating this belief among the peasantry as a way of protecting wolf dens until the fall hunting season, 1524.
167 A. Ventseslavsky, "Zaokhtenskie volki," *Zhurnal konnozavodstva i okhoty* 11 (November 1854): 96.
168 Ibid.
169 Ibid., 105.
170 Ibid., 100.
171 Sabaneev, *Volk*, 1406–15. I am presenting only the highlights of Sabaneev's analysis.
172 Ibid., 1417–19.
173 Ibid., 1416.
174 Ibid., 1428–29.
175 Ibid., 1430–35.
176 Ibid., 1446.
177 Ibid., 1446–47.
178 Ibid., 1449.
179 Ibid. A thorough description of the sorts of economic changes that Sabaneev had in mind can be found in Thomas C. Owen, *The Corporation Under Russian Law, 1800-1917: A Study in Tsarist Economic Policy* (Cambridge: Cambridge University Press, 1991).
180 It's ironic that Sabaneev couched this lament for borzoi hunting in such traditionalist terms, given that he himself—educated in natural history of Moscow State University in the early 1860s and in light of his earlier criticism of the exploitation of the serfs prior to emancipation—was a member of the new generation, although from a noble family.
181 Sabaneev, *Volk*, , 1463.
182 Ibid., 1473.
183 Ibid., 1525.
184 Ibid., 1523.
185 Ibid., 1525.
186 N. V. Turkin, *Zakony ob okhote*. This is the editorial which Cavender mistakenly ascribes to Sabaneev.
187 Turkin, *Zakon ob okhote 3 fevralia 1892 goda* (Moscow: Tipografiia M. G. Volchaninova, 1892).
188 Turkin, *Okhota i okhotnich'e zakonodatel'stvo*.
189 Turkin, *Zakony ob okhote*, ii–iii.
190 Ibid., 18.

191 Ibid., 103–9.
192 Ibid., 93.
193 Ibid., 6–7, 13.
194 Ibid., 70.
195 Ibid., 68.
196 Ibid., 21–22.
197 Ibid., 71–72.
198 Ibid., 75–76.
199 Ibid. For an in-depth discussion of Sweden's anti-wolf policies focusing on the eighteenth century which corroborates Turkin's analysis see Roger Bergstrom, Karin Dirke, and Kjell Danell, "The Wolf War in Sweden during the Eighteenth Century—Strategies, Measures, and Leaders," in Masius and Sprenger, *A Fairytale in Question*, 57–78. For a detailed discussion of wolf-reduction efforts in Germany in the eighteenth and nineteenth centuries see Patrick Masius and Jana Sprenger, "Reconstructing the Extermination of Wolves in Germany: Case Studies from Brandenburg and Rhineland-Palatinate," in Masius and Sprenger, *A Fairytale in Question*, 119–40.
200 Turkin, *Zakony ob okhote*, 76–79. Turkin was a meticulous researcher, but such exact numbers are inherently problematic in the context of his era. For helpful discussions of the reliability of journalistic accounts and other sources of information about wolf attacks in earlier eras see Karin Dirke, "Where Is the Big Bad Wolf? Notes and Narratives on Wolves in Swedish Newspapers during the 18th and 19th Centuries," in Masius and Sprenger, *A Fairytale in Question*, 101–118; and Linnell et al., eds., *The Fear of Wolves*, 8–13. The international and highly collaborative Linnell study suggests that "scientific, veterinary, and medical" accounts, as well as "historical and administrative records" such as parish registers provide the most reliable sources (8). It suggests that caution be used in drawing conclusions from "newspapers, non-technical literature and from interviews or personal communications" (8). Throughout this book I have tried to err on the side of caution in interpreting journalistic accounts while also recognizing the valuable insights they provide.
201 Turkin, *Zakony ob okhote*, 79–80.
202 Ibid., 81.
203 For the text of the law see *Polnoe sobranie zakonov Rossiiskoi Imperii*, ser. 3, vol. 12 (1892), no. 8301, 81–85. Given its relative brevity, I will not cite a location for each of the points outlined below.
204 Ibid., 83.
205 Ibid.
206 Turkin, *Zakon ob okhote 3 fevralia 1892 goda*, 68–77, contains Turkin's summary of the considerations that went into the portion of the law relating to predators.
207 I will delve into modern perspectives on the role of apex predators and related issues in the conclusion. While I have cited Turkin's 1889 book much more extensively for practical reasons, his elucidation of the 1892 law is strikingly well-informed and detailed. It proceeds sequentially through each of its articles, providing context and summarizing the ministerial-level discussions that led to each of them, citing the perspectives of individual ministers and other members of the commission that had been established in 1889 under the auspices of the Ministry of State Properties to craft the new law. For his explanation of the genesis of this commission, see Ibid., "Predislovie," i–iv.
208 Pravilova, *A Public Empire*, 71.
209 Turkin, "Istoricheskii ocherk," 58.
210 Turkin, *Okhota i okhotnich'e zakonodatel'stvo*, 146.

211 Ibid., 144.
212 The story has been reprinted as I. A. Salov, "Volki," in *Russkii okhotnichy rasskaz*, comp. M. M. Odesskaya and ed. N. I. Netesnina (Moscow: Sovetskaya Rossiia, 1991), 163–78. I have not been able to pinpoint its original year of publication. The 1991 edition draws on Salov's collected works, published posthumously in 1908 to 1910, as its source. For an overview of Salov's life and work see P. V. Bykov, "I. A. Salov," in I. A. Salov, *Sochineniia* (St. Petersburg: Tip. A. F. Marks, 1908), 5: 3–27. See also Odesskaya, "Primechaniia," in *Russky okhotnichy rasskaz*, 422–23.
213 Ibid., 165.
214 Ibid., 165–66.
215 Ibid., 170.
216 Ibid., 171.
217 Ibid.
218 Ibid., 172.
219 Ibid., 177.

NOTES FOR CHAPTER 3

220 A. Kh., "O beshenstve u liudei i zhivotnykh," *Priroda i okhota*, January 1880, 16.
221 For a wide-ranging overview of the disease in human history see Bill Murphy and Monica Wasik, *Rabid: A Cultural History of the World's Most Diabolical Virus* (New York: Viking Penguin, 2012). For an exploration of rabies in the very different context of Britain starting in the mid-nineteenth century see Neil Pemberton and Michael Worboys, *Rabies in Britain: Dogs, Disease and Culture 1830-2000* (Basingstoke, UK: Palgrave Macmillan, 2012).
222 In his 1889 study of hunting law, for example, Turkin compiled a list of thirty-eight cases between 1870 and 1879 in which the press had reported attacks on people by rabid wolves, mostly within European Russia: Turkin, *Zakony ob okhote*, 77–79.
223 One of the first journalistic accounts of their visit was submitted to the *Moscow Gazette* by an unnamed Parisian doctor who'd spent time in Russia: "Russkie bol'nye krest'iane v Parizhe," *Moskovskie vedomosti*, March 23, 1886, 5. For a scholarly description of the episode see A. S. Shevelev, "L. Paster i smoliane (perepiska Pastera)," *Sovetskoe zdravookhranenie* 3 (1979): 56–59.
224 A. P. Chekhov, "Volk," *Polnoe sobranie sochinenii i pisem v tridtsati tomakh*, ed. N. F. Bel'chikov (Moscow: Nauka, 1986), 5: 39–45 (text), 494–99 (variations), 616–17 (notes). The terms [*sobach'e*] *beshenstvo* and *vodoboiazn'* were both used to describe rabies during the imperial period, although the former was considered more scientific and gradually superseded the latter.
225 T. A. Kuzminskaya, "Beshenyi volk. Istinnoe proisshestvie," *Vestnik Evropy* 6 (June 1886): 595–612.
226 Charles E. Rosenberg, "Framing Disease: Illness, Society, and History," in C. E. Rosenberg and Janet Golden, *Framing Disease: Studies in Cultural History* (New Brunswick, NJ: Rutgers University Press, 1992), xiii–xxvi.
227 See Michel Foucault, *History of Madness in the Classical Age*, ed. Jean Khalfa and trans. Jonathan Murphy and Jean Khalfa (London: Routledge, 2006) for a compilation of Foucault's wide-ranging investigations into the ways in which societies have defined and institutionalized madness.
228 "Strashnyi sluchai" (see n. 1). No author was listed for the article, but the anonymous account closely reflects the views of the medical professionals who cared for the patients and was likely authored or based closely on an account by Grabovsky.

229 Ibid., 102-3.
230 Ibid., 104.
231 For a minutely researched description of European rabies treatments in this era see K. Codell Carter, "Nineteenth-Century Treatments for Rabies as Reported in the *Lancet*," *Medical History* 26 (1982): 67-78. For a synopsis of 693 cases from the 1820s through the 1880s of patients who had been admitted for possible rabies exposure and their subsequent treatments at one of Moscow's major hospitals for the poor see D. P. Kishensky, "K voprosu o vodoboiazni (po arkhivnym dokumentam Moskovskoi Ekaterin in skoi bol'nitsy)," *Vrach* 44 (October 29, 1887): 849-52, and 45 (November 5, 1887): 871-72. I will return to Kishensky's analysis below. *Vrach*, established in 1880, was Russia's leading medical journal and one to which Chekhov subscribed (according to Mondry, *Political Animals*, 100).
232 Rosenberg, xvi.
233 Danilo Samoilovich, *Nyneshny sposob lecheniia s nastavleniem kak mozhno prostomu narodu lechit'sia ot ugryzeniia beshenoi sobaki i ot uiazvleniia zmei*, 2nd ed. (Moscow: Tipografiia u N. Novikova, 1783).
234 Ibid., 12-13.
235 Ibid., 13-14. Samoilovich cites the sixteenth-century Leipzig theologian Heinrich Salmuth as the source for this presumably apocryphal story.
236 Physician Juan Gomez-Alonso has argued that rabies may provide an explanation for these creatures of folklore and their close association with dogs and wolves. He traces the similarities between the symptoms of rabies—including restlessness, hypersensitivity, spasms, aerophobia, facial grimaces, etc.—and those ascribed to both vampires and werewolves in a range of eighteenth-century accounts. See Juan Gomez-Alonso, "Rabies: A Possible Explanation for the Vampire Legend," *Historical Neurology* 51 (1998): 856-59.
237 Ibid., 14-15.
238 Ibid., 29-31, 36-47.
239 M. P. Maroketti, *Praktichesky i teoretichesky traktat o vodoboiazni, soderzhashchy v sebe predokhranitel'nuiu metodu ot beshenstva*, 2 vols. (St. Petersburg: Tipografiia Ministerstva vnutrennikh del, 1840).
240 Ibid., 1: 15-16.
241 Ibid., 1: 26-27. In general, cauterizing wounds and/or applying caustic compresses to them were prevalent throughout the nineteenth century based on the theory that this might prevent the rabies "poison" from penetrating into the organism.
242 N. A. Arendt, *Vodoboiazn', pes'e beshenstvo* (Simferopol': Gubernskaya tipografiia, 1859).
243 Ibid., 3-5.
244 Ibid., 7-9.
245 Ibid., 32.
246 Pasteur's vaccine, despite its efficacy, predated scientific recognition of viruses. For an efficient description of the discovery of viruses at the onset of the twentieth century, including the RNA virus that causes rabies, see David Quammen, *Spillover: Animal Infections and the Next Human Pandemic* (New York: Norton, 2012): 263-72. The rabies virus travels along the axoplasm, which conducts electrical impulses through the nervous system to and from the brain, at an average rate of one to two centimeters a day. For this reason severe bites closest to the head—those typically inflicted by wolves—pose the greatest danger. See also Linnell et al., eds., *Fear of Wolves*, 14, for an efficient overview of the transmission and typical course of the disease.
247 Carter, "Nineteenth-Century Treatments for Rabies," makes this point very convincingly in relation to changing attitudes toward rabies treatments in Western Europe, 76-77.

Speaking more generally, Rosenberg notes: "In earlier centuries lay and medical views of disease overlapped to some extent, so that shared knowledge tended to structure and mediate interactions between doctors, patients, and families" (xviii). *Vrach* provided frequent updates to Pasteur's experiments starting in 1882 and culminating in 1885 to 1886, while also publishing occasional articles advocating other approaches to treating rabies; these dwindled rapidly after Pasteur's discovery.

248 A. Kh., "O beshenstve u liudei i zhivotnykh," 3.
249 Ibid., 9–10.
250 Ibid., 10–12.
251 Aside from the brief editorial notes in Chekhov's collected works, the story appears to have received almost no scholarly attention. A. Yarmolinsky translated it as "Hydrophobia" in his *The Unknown Chekhov: Stories and Other Writings Hitherto Untranslated* (New York: Noonday, 1954), 95–106. He provided both the earlier and later endings as well as one other variant passage and devoted two paragraphs to it in his general introduction, 14–15. He did not, however, attempt to analyze the relationship between the two versions in detail, interpret the role of the doctor(s), nor provide any context concerning wolves or rabies in Chekov's era. In fact, his assertion that Chekhov "made some minor changes in the wording, deleted two short passages, and completely altered the ending" does not do justice to the significant differences between the two versions in addition to the changed ending, 95.
252 Chekhov, *PSS*, 5: 39; 494. Mondry rightly emphasizes that the Russian word *beshennyi* (which means mad or crazy, but also rabid) is semantically linked to the word *besy* (devil or possessed), so that "a mad dog is also a possessed dog, a concept that denotes not only a medical conditional but also a supra-natural state" (*Political Animals*, 132). This helps to contextualize Maksim's fears.
253 Chekhov, *PSS*, 5: 40; 494–95.
254 Ibid., 40.
255 A provincial doctor named Ovchinnikov also features in Chekhov's 1888 story "An Unpleasantness," which centers on the doctor's sense of impotence at his inability to deal appropriately with a drunken assistant: Chekhov, "Nepriatnost'," *PSS*, 7: 141–58.
256 Chekhov, *PSS*, 5: 39–40.
257 Ibid., 40–41, 494–95.
258 Ibid., 41, 496. Chekhov referred to this passage in a letter of May 10, 1886 to his brother Alexander, affirming that nature description should be simple and spare: "In my opinion descriptions of nature must be very brief and possess an *à propos* quality. In describing nature you should grasp on to the slightest details, grouping them in such a way that you can then close your eyes and see a picture. For example, you can create a moonlit night if you write that on a millhouse weir a piece of glass from a broken bottle sparkled like a star and the dark shadow of a dog or wolf moved in a circle, etc. Nature becomes animated if you don't disdain comparing its phenomena with human actions, etc." (A. P. Chekhov, *PSS, Pis'ma*, 1: 242). Yarmolinsky has also noted the connection between letter and passage, 15.
259 Jane Costlow quotes the story's climax, emphasizing the "terrifying otherness and massive strength" of the bear, as well as the hunter's ignorance as a tracker in comparison with his professional guide in "For the Bear to Come to your Threshold: Human-Bear Encounters in Late Imperial Russian Writing," *Other Animals*, 92.
260 Chekhov, *PSS*, 5: 496.
261 Ibid., 42.
262 Ibid., 44–45, 498.
263 Kishensky, "K voprosu o vodoboiazni," 44: 50.
264 Chekhov, *PSS*, 7: 141.

265 Ibid., 5: 499.
266 While Pasteur's vaccine is mentioned only briefly in the 1886 version of the story, Chekhov published "To Paris" (*V Parizh*) just five days later under the same early pseudonym in the journal *Fragments (Oskolki)*. Written in the comic vein that characterized many of his early works, the story describes events in a provincial town after a rabid dog bites a minor official and schoolteacher. It culminates in their abortive trip to seek treatment directly from Pasteur, which comes to an early end when they disembark from the train while still in Russia and drink away the travel money that has been raised by their fellow townspeople. While Chekhov edited "Hydrophobia" significantly in the 1890s, demonstrating his serious interest in the work, he specifically and emphatically noted that "V Parizh" should not be included in his collected works. For the story and scholarly annotations see Chekhov, *PSS*, 5: 46–51, 617–18.
267 Tolstoy, *PSS*, 63: 288–89. Tolstoy also praised the work in an October 16 letter to his wife, noting that it "reads easily and with the interest of horror" (Tolstoy, *PSS*, 83: 513–14). It's unclear whether he followed through with his intention of editing the story. It has received almost no attention, unlike Kuzminskaya's memoirs and her better-known story—also drawn from a real-life narrative—about an unfortunate peasant woman entitled "A Woman's Lot," which also appeared in the *European Herald* in 1886. For insight into Tolstoy's editorial relationship with Kuzminskaya see Z. N. Ivanova, "T. A. Kuzminskaya: Bab'ia dolia (Rasskaz krest'ianki)," in *Tolstoy-redaktor*, ed. E. Ie. Zaidenshnur, 77–87 (Moscow: Kniga, 1965).
268 Kuzminskaya, "Beshenyi volk," 599.
269 Ibid., 601.
270 Ibid., 602. She uses a non-medical term, *shalyi*, in suggesting the wolf isn't rabid (*beshennyi*).
271 Ibid.
272 Ibid., 605.
273 Ibid., 607.
274 Ibid., 608. The term implies the use of incantations or prayers for healing. In the 1862 account cited earlier the term used pejoratively to describe the exploitative healers is *znakhar'*, which can be translated as "healer" but also as "quack."
275 Ibid., 608–9.
276 Kishensky's archival analysis of treatments from the 1820s to the 1870s at Moscow's Ekaterininskaya Hospital, cited earlier, reflects the institutional practices that may have helped to inform Kuzminskaya's portrayal. As in her narrative, the hospital typically kept patients for six weeks to ascertain whether they would develop rabies. Treatments included cold water baths that rendered patients unconscious, cauterization with hot metal or caustic agents, excision of the wounded flesh, internal agents including both mercury and arsenic, and a variety of other approaches (Kishensky, "K voprosu o vodoboiazni," 44: 851–52). One of the most striking case studies Kishensky found in the archives involved a house serf in her early sixties who was sent to the hospital after her mistress had her lapdog bite her "to determine whether it had rabies" and who died amid a series of cold water treatments (Kishensky, 45: 871–72). Kuzminskaya's narrative does not delve into the methods employed by the provincial hospital, but the victims' brief encounter with institutionalized medicine echoes the tenor of these practices.
277 Kuzminskaya, "Beshenyi volk," 609.
278 Ibid.
279 Ibid., 612.
280 Ibid.

281 In addition to the sources cited below I am drawing the facts of the Smolensk story from those mentioned in note 223.
282 My perusal of *Vrach* throughout the 1880s revealed dozens of articles and briefer references to Pasteur's rabies experiments, the earliest in 1883, with an especially dense concentration in 1885 to 1887.
283 A. S. Shevelev, "O roli S. A. Rachinskogo v organizatsii poezdki zhitelei smolenskoi gubernii k Pastery," *Zhurnal mikrobiologii, epidemiologii i immunobiologii* 5 (1979): 113–14, draws on archival sources to explicate Pobedonostsev's involvement, which arose through energetic efforts on the peasants' behalf by S. A. Rachinsky (a Moscow University professor of botany and the first translator of Darwin into Russian), 114.
284 "Russkie bol'nye krest'iane v Parizhe," 5.
285 Ibid.
286 Ibid.
287 L. Paster [Pasteur], *Beshenstvo*, trans. N. B. Lanchulizdeva and M. O. Perel'man (Saratov, Russia: Saratovsky Uezdnyi Zemsky Sanitarnyi Sovet, 1891), 13.
288 Ibid.
289 Ibid.
290 The most detailed account of the process that led to the establishment of Russia's early rabies stations that I've encountered, in addition to updates that appeared in *Vrach* at the time, is E. V. Sherstneva, "Pervye Pasterovskie stantsii v Rossii," *Problemy sotsial'noi gigieny, zdravookhraneniia, i istorii meditsiny* 3–4 (2012): 56–59. For a first-hand account by a Russian medical researcher who had traveled to Paris to train with Pasteur and then helped to establish the Petersburg rabies station see N. A. Kruglevsky, *O privivke liudiam iada sobach'ego beshenstva po sposobu Pastera* (St. Petersburg: Tipografiia Ia. Trei, 1887). Historian of science Daniel Philip Todes also devotes several pages at the beginning of his *Pavlov's Physiology Factory: Experiment, Interpretation, Laboratory Enterprise* (Baltimore: Johns Hopkins University Press, 2002), to discussing the events that led to the establishment of rabies stations in Russia as well as the Petersburg research institute subsequently directed by Pavlov; he focuses on the role of the well-connected aristocrat, Prince A. P. Ol'denburgsky, in promoting and supporting them, 4–10.
291 Sherstneva, "Pervye Pasterovskie stantsii," 58, provides the figures on which I have calculated the number of more than 100,000 patients. The Odessa station alone treated 47,564 patients between 1887 and 1917, according to her statistics.
292 To give just two examples see K. A. Bari, *Beshenstvo*, 3rd ed. (Moscow: Institut Pastera pri Bol'nitse im. Imperatora Aleksandra III, 1912); and [Vrach] Kartashevsky, *Samaya strashnaya bolezn' (Beshenstvo). Dlia krest'ian*, ed. Iu. I. D'iakov (Moscow: Novaya Moskva, 1926).

NOTES FOR CHAPTER 4

293 S. Pospelov, "Zhestokaya zabava. Travlia," *Zashchita zhivotnykh* 11 (November 1905): 450–51.
294 For a comprehensive history of the first ten years of the Society's existence, see V. Iversen, *Pervoe desiatiletie Rossiiskago Obshchestva pokrovitel'stva zhivotnym. Istorichesky ocherk ego deiatel'nosti v 1865-1875 gg.* (St. Petersburg: Tipografiia A. M. Kotimina, 1875).
295 For a helpful overview of attempts to protect bird populations in late imperial Russia, see Brian Bonhomme, "For the 'Preservation of Friends' and the 'Destruction of Enemies':

Studying and Protecting Birds in Late Imperial Russia," *Environment and History* 13, no. 1 (2007): 71–100.
296 See Alexandra Popoff, *Tolstoy's False Disciple: The Untold Story of Leo Tolstoy and Vladimir Chertkov* (New York: Pegasus, 2014) for a thorough discussion of the fascinating relationship between the two men that draws on a wide variety of sources, including previously neglected archives. I have drawn on her work for some of the basic biographical and other details in this paragraph.
297 Tolstoy, *PSS*, 87: 49.
298 Ibid., 48. I've translated both instances of the verb *popravliat'* as "correct." Tolstoy often used this verb rather than the more typical forms corresponding with "edit" or "improve."
299 Ibid., 50–51.
300 Ibid., 52.
301 Tolstoy, "Predislovie" [k stat'e V. Ch-va, "Zlaya zabava"], *Novoe vremia* 5284 (November 13, 1890): 2.
302 V. Ch-v, "Zlaya zabava," 3.
303 Ibid.
304 Ibid.
305 Ibid. Less than a year after the publication of the article Chertkov provided Tolstoy with a copy of the English writer and advocate of vegetarianism Howard Williams' *The Ethics of Diet*, for the Russian translation of which Tolstoy wrote an introduction that was published separately in 1892 as "The First Step." See R. F. Christian, "Tolstoy and the First Step," *Scottish Slavonic Review* 20 (1993): 7–16. See also note 57.
306 "Znachenie okhoty (po povodu zametki grafa L. N. Tolstago i stat'i g. Ch-va)," *Okhotnich'ia gazeta*, November 26, 1890, 738.
307 Ibid.
308 Ibid.
309 M. F. Aderkas, "Zlaya zabava," *Vestnik Rossiiskago Obshchestva pokrovitel'stva zhivotnym*, January 1891, 15–17.
310 Ibid., 17. For a more detailed discussion of this episode see Iversen, *Pervoe desiatiletie Rossiiskago Obshchestva pokrovitel'stva zhivotnym*, 31–36.
311 Amy Nelson provides an excellent summary of the society's history from its founding to its demise on the eve of the revolutionary era. She pays particular attention to its significant but constrained efforts to effect meaningful animal protection legislation and to gain police cooperation in enforcing anti-cruelty measures, noting that it provided a poignant example of the ways in which Russia's emerging civil society struggled to gain a meaningful voice and role within the confines of the tsarist regime and bureaucracy. The RSPA was under the jurisdiction of the Ministry of Internal Affairs. Although members of the society and especially those designated as "wardens" could report infractions, they relied on local police to enforce anti-cruelty measures. See Amy Nelson, "The Body of the Beast: Animal Protection and Anticruelty Legislation in Imperial Russia," in Costlow and Nelson, *Other Animals*, 95–112.
312 Iversen, *Pervoe desiatiletie*, 31–38.
313 Ibid., 36–37. The government lent its support to this prohibition in 1867 but imposed a five-year wait on implementation in order to minimize the disruption for those who made their livelihood from the practice. Nelson also discusses this in "Body of the Beast," 101.
314 G. Voronov, "Zversky i opasny sposob istrebleniia volkov," *Vestnik Rossiiskago Obshchestva pokrovitel'stva zhivotnym*, June 1893, 207–8. Reports of this method had appeared as early as 1877 in the *Forestry Journal*: "Original'nyi sposob istrebleniia volkov," *Lesnoi zhurnal* 5 (1877): 89.

315 Voronov, "Zversky i opasny sposob," 207.
316 Ibid., 208.
317 *Sadka* derives from the verb *sazhat'* (to seat, set, put, plant) and has a much more neutral ring in Russian than does *travlia*. Its use as a hunting term is archaic. Dahl's historical dictionary, first published in 1863 to 1866 and a standard reference for such terms, uses *travlia* as a synonym to define the practice (V. Dahl, *Tolkovyi slovar' zhivago velikorusskago iazyka*, ed. V. V. Pchelkina, 4 vols. [Moscow: Russky iazyk, 1978-80], 4: 128). Because animal "baiting" has a wider set of meanings, I will generally use the term "hounding" in my own discussion.
318 V. S. Tolstoy, "Sadka na volkov 4 ianvaria 1880 g.," *Priroda i khota*, February 1880, 263. Tolstoy concluded his account by describing the enclosures in which the wolves were confined and from which they were released to the borzois. He noted that they had been artfully designed by "the well-known borzoi hunter and wolf-hounder N. V. M." in such a way as to expose the wolf on all four sides at once on its release (V. S. Tolstoy, "Sadka," 265–66). As in the accounts we explored earlier, the name "Ubei" reflects ruthlessness, as it corresponds to the imperative form of the verb *ubit'* (to kill).
319 Chekhov, "Na volch'ei sadke," *PSS*, 1: 117–121 (text), 572–73 (notes).
320 Ibid., 117.
321 Ibid., 118.
322 Ibid., 119. All ellipses in this quote are Chekhov's own.
323 Ibid., 119.
324 Ibid., 120.
325 Ibid., 118.
326 Z-o-i Z-o-i, "Travli zhivotnykh," *Vestnik Rossiiskago Obshchestva pokrovitel'stva zhivotnym* 2 (February 1890): 31.
327 A. N. Kremlev, "Sadki na rezvost' i zlobu. Doklad pravleniiu Rossiiskago Obshchestva pokrovitel'stva zhivotnym," *Vestnik Rossiiskago Obshchestva pokrovitel'stva zhivotnym* 6 (June 1894): 172.
328 Ibid., 170–71.
329 Ibid., 175.
330 S. Pospelov, "Zhestokaya zabava. Travlia," *Zashchita zhivotnykh* 11 & 12 (November and December 1905): 441–53; 491–95. I will discuss Pospelov in more detail below.
331 Ibid., 11: 450–51.
332 S. Pospelov, "Dva brata (Iz volch'ei zhizni), in *Razskazy o dikikh zhivotnykh: Knizhka pervaya*, 2nd edition (St. Petersburg: Tipografiia O. N. Popovoi, 1914), 13–33.
333 Ibid., 28.
334 Sabaneev, *Volk*, 1419.
335 E. Chernobaev, "S. Pospelov. Razskazy o dikikh zhivotnykh Rossii," *Vestnik Rossiiskago Obshchestva pokrovitel'stva zhivotnym*, 1 (January 1906): 27.
336 Matt Cartmill traces the polemics that pitted Seton and his fellow Canadian nature writer William J. Long against the more hard-headed writer John Burroughs, who famously accused the two in a 1903 *Atlantic* article of depicting animals unrealistically in pursuit of public favor and profit. See his *A View to a Death in the Morning: Hunting and Nature Through History* (Cambridge, MA: Harvard University Press, 1993), 150–56. Seton had worked as a government bounty hunter earlier in his life and depicts his own killing of a wolf in the story "Lobo," which is included in his collection *Wild Animals I Have Known*.
337 A. L., "Zhizn' i prikliucheniia odnogo volka," *Priroda i okhota*, April 1892, 58–91.
338 Ibid., 68.
339 Ibid., 68–69.

340 Ibid., 73.
341 Ibid., 71.
342 Ibid., 74.
343 Ibid., 77.
344 Ibid., 78.
345 Ibid., 82.
346 Ibid., 91.
347 A. P. Chekhov, "Belolobyi," *PSS*, 9: 100–106 (text), 394 (variants), 467–69 (notes). My comments on the story's publication history are based on the editorial notes in the scholarly edition.
348 Ibid., 469.
349 Ibid., 100.
350 Ibid., 102.
351 Ibid., 103.
352 Ibid., 104.
353 For an overview of Zaitsev's life and work see Odesskaya, "Primechaniia," in *Russky okhotnichy rasskaz*, 422. For a substantially more detailed account see A. M. Berezkin, "Zaitsev, Boris Konstantinovich," in *Russkie pisateli 1800-1917. Biograficheksy slovar'*, ed. P. A. Nikolaev (Moscow: FIANIT, 1992), 2: 309–313. My brief encapsulation of Zaitsev's biography and publications is based on these two sources.
354 Berezkin, "Zaitsev," 312.
355 Odesskaya, "Primechaniia," 422.
356 The story has been reprinted as B. K. Zaitsev, "Volki," in *Russky okhotnichy rasskaz*, 158–62, utilizing a 1905 collection of stories by Zaitsev as its source.
357 Ibid., 158.
358 Ibid.
359 Ibid.
360 Ibid., 159.
361 Ibid., 160.
362 Ibid.
363 Ibid.
364 Ibid.
365 Ibid.
366 Ibid., 161.
367 Ibid.
368 Ibid., 162. In 1917, the writer Boris Pil'niak wrote a story entitled "Pozemka" ("The Snow Wind") that is very reminiscent of Zaitsev's earlier narrative. I have chosen not to include it as the 1902 story contributes more readily to my overall analysis in this chapter. See B. Pil'niak, "Pozemka" in *Byl'e* (Revel', Russia: Bibliofil, 1922): 104–112.
369 See Jane Costlow, "The Gallop, The Wolf, The Caress: Eros and Nature in *The Tragic Menagerie*," *The Russian Review* 56 (April 1997): 192–208. Given Costlow's superb translation, I will cite directly from it: L. D. Zinovieva-Annibal, *The Tragic Menagerie*, trans. and introd. Jane Costlow (Evanston, IL: Northwestern University Press, 1999). For the original see L. D. Zinovieva-Annibal, *Tragichesky zverinets* (St. Petersburg: Ory, 1907).
370 Costlow, "The Gallop," 193.
371 Zinovieva-Annibal, 21.
372 Ibid.
373 Ibid.

374 As Costlow notes, "[t]he hunt in this story is associated not with the need for sustenance or protection, but with spectacle and diversion" ("The Gallop," 199).
375 Zinovieva-Annibal, 26.
376 Ibid., 27.
377 Ibid., 27–28.
378 Ibid., 28–29.
379 Ibid., 29.
380 Ibid., 36.
381 See Costlow's probing interpretation of Vera's conversations with Fyodor and her mother, which overlaps with mine but emphasizes different aspects, in "The Gallop," 201–3. Costlow emphasizes that Vera's feelings of "kinship" with the wolf reflect her own maimed desire for freedom.
382 Catherine, Elick, *Talking Animals in Children's Fiction: A Critical Study* (Jefferson, NC: McFarland & Co., 2015), 9.

NOTES FOR CONCLUSION

383 Aldo Leopold, *A Sand County Almanac* (New York: Ballantine, 1966), 138–39. For an exhaustive historical analysis of the major Western tendencies and figures that set the stage for twentieth-century environmentalism, including Leopold, see Donald Worster, *Nature's Economy: A History of Ecological Ideas*, 2nd ed. (Cambridge: Cambridge University Press, 1994). Douglas Weiner, *Models of Nature: Ecology, Conservation, and Cultural Revolution in Soviet Russia* (Bloomington: Indiana University Press, 1988) presents an overview of nascent conservationism in late tsarist Russia, 7–18, and devotes the body of his book to tracing it through the Soviet period.
384 Andrew C. Isenberg, "The Moral Ecology of Wildlife," in *Representing Animals*, ed. Nigel Rothfels (Bloomington: Indiana University Press, 2002): 54.
385 For a concise overview of the late twentieth-century evolution among ecologists to emphasizing the importance of apex predators and the related concept of "keystone species," whose presence helps to maintain an ecosystem's stability, see Quammen, *Monster of God*, 416–24.
386 Isenberg, 51–55.
387 Coleman discusses this famous passage and the larger contours of wolf eradication efforts in the final chapter of *Vicious*, 191–224. See also Rick McIntyre, *War Against the Wolf: America's Campaign to Exterminate the Wolf* (Stillwater, MN: Voyageur Press, 1995).
388 See note 36.
389 See note 100.
390 See note 260.
391 See note 331.
392 Armstrong, "The Gaze of Animals," 181 (see note 31).
393 Ibid., 188.
394 Armstrong cites S. Freud's celebrated case study of the "Wolf Man" (1918) in this portion of his argument, 189. According to Freud's interpretation, the Russian nobleman Sergei Pankeev's dream of seven white wolves staring at him through his bedroom window stemmed from a repressed memory of seeing his parents copulating as an infant, so that the wolves' gaze actually represented Pankeev's repressed memory of himself staring at his father, transformed into a group of wolves. This interpretation

undermines the power of the animal gaze, appropriating it instead to explore the troubled psyche of the human subject. For a translation of Freud's case study with further analysis and context see Muriel Gardiner, ed., *The Wolf-Man and Sigmund Freud* (London: Karnac Books, 1989).

395 Armstrong touches on some of the highlights of Derrrida's erudite and nuanced thinking on this topic. For the original essay of well over 100 pages see Jacques Derrida, *The Animal That Therefore I Am*, ed. Marie-Louise Mallet, trans. David Wills (New York: Fordham University Press, 2008).

396 Armstrong, 193–97. For a book-length exploration of the animal gaze in a different geographical and cultural context see Wendy Woodward, *The Animal Gaze: Animal Subjectivities in South African Narratives* (Johannesburg, South Africa: Wits University Press, 2008).

397 See my discussion of Graves and Rathgeber in notes 20 and 21. Among the studies of wolf reintroduction efforts in the Yellowstone region see Tim W. Clark, Murray B. Rutherford, and Denise Casey, eds., *Coexisting with Large Carnivores: Lessons from Greater Yellowstone* (Washington, DC: Island Press, 2005). For a recent exploration of the Yellowstone wolf reintroduction that is structured around the story of one of the most famous individuals in the Druid pack of the Lamar Valley see Nate Blakeslee, *American Wolf: A True Story of Survival and Obsession in the West* (New York: Crown [Penguin], 2017). For a collection of perspectives on the potential to reintroduce wolves in the northeastern United States see John Elder, ed., *Return of the Wolf: Reflections on the Future of Wolves in the Northeast* (Hanover, NH: Middlebury College Press, 2000). See Marco Musiani, Luigi Boitani, and Paul C. Paquet, eds., *A New Era for Wolves and People: Wolf Recovery, Human Attitudes, and Policy* (Calgary: University of Calgary Press, 2009) for analysis of current reintroduction and recovery efforts and prospects for human-wolf coexistence in both North America and Europe.

398 See Hans Kruuk, *Hunter and Hunted: Relationships between Carnivores and People* (Cambridge: Cambridge University Press, 2002), 68–73. A zoologist, Kruuk highlights this contrast but offers no ready explanation. See also Linnell, *Fear of Wolves*, 24–28, and notes 12 and 19.

399 The passage can be found in F. M. Dostoevsky, *Polnoe sobranie sochinenii v 30 tomakh*, ed. V. G. Bazanov (Leningrad: Nauka, 1976), 15: 221. The two journals to which Ivan refers—*Russky arkhiv* and *Russkaya starina*—often printed material from the late eighteenth and early nineteenth centuries. The story corresponds to a passage in "Memoirs of a Serf," which appeared in 1877 in yet another journal, *Russky vestnik* (see ibid., 554). For a translated version see Fyodor Dostoevsky, *The Brothers Karamazov*, translated by Richard Pevear and Larissa Volokhonsky (New York: Vintage, 1990), 242–43.

400 Mondry provides an insightful reading of this scene in her *Political Animals*, 47–48. She rightly points out that the version in *Russky vestnik* states that the borzois "would not touch him" when unleashed on the boy for the first time and likely did not even when released on him again, as borzois were trained not to attack people and to hold rather than tear apart. In other words, Ivan's reinvention of the story diverges both from its likely source and the norms of borzoi hunting. I agree with her interpretation yet find Dostoevsky's creative reconfiguration of the historical patterns we have explored illuminating.

401 Graves helpfully includes a list of Russian proverbs about wolves in his *Anxiety through the Ages*, 141–70.

Bibliography

Aderkas, M. F. "Zlaya zabava." *Vestnik Rossiiskago Obshchestva pokrovitel'stva zhivotnym*, January 1891, 15–17.

Arendt, N. A. *Vodoboiazn', pes'e beshenstvo.* Simferopol': Gubernskaya tipografiia, 1859.

Armstrong, Phillip. "The Gaze of Animals." In *Theorizing Animals: Re-Thinking Humanimal Relations*, edited by Nik Taylor and Tania Signal, 175–99. Leiden, The Netherlands: Brill 2011.

Bari, K. A. *Beshenstvo*, 3rd ed. Moscow: Instititut Pastera pri Bol'nitse im. Imperatora Aleksandra III, 1912.

Berger, John. "Why Look at Animals?." In *About Looking*, 3–28. New York: Pantheon, 1980.

Belov, Vlad. "K volch'emu voprosu." *Priroda i okhota*, January 1878, 49–51.

Berezkin, A. M. "Zaitsev, Boris Konstantinovich." In *Russkie pisateli 1800-1917. Biograficheksy slovar'*, 2: 309–13. Edited by P. A. Nikolaev. Moscow: FIANIT, 1992.

Bergstrom, Roger, Karin Dirke, and Kjell Danell. "The Wolf War in Sweden during the Eighteenth Century—Strategies, Measures, and Leaders." In *A Fairytale in Question: Historical Interactions between Humans and Wolves*, edited by Patrick Masius and Jana Sprenger, 57–78. Cambridge, UK: The White Horse Press, 2015.

Bezobrazov, S. "Ocherki lesnoi okhoty v Iamburgskom uezde." *Priroda i okhota*, April 1878, 17–28; May 1878, 113–24.

Bibikov, D. I., ed. *Volk: Proiskhozhdenie, sistematika, morfologiia, ekologiia.* Moscow: Nauka, 1985.

———, ed. *Ekologiia, povedenie i upravlenie populiatsiiami volka. Sbornik nauchnykh trudov.* Moscow: VASKhNIL, 1989.

———, ed. *Der Wolf: Canis Lupus.* Translated by Günther Grempe. Wittenberg Lutherstadt: A. Ziemsen, 1988.

Blakeslee, Nate. *American Wolf: A True Story of Survival and Obsession in the West.* New York: Crown [Penguin], 2017.

Bonhomme, Brian. "For the 'Preservation of Friends' and the 'Destruction of Enemies': Studying and Protecting Birds in Late Imperial Russia." *Environment and History* 13, no. 1 (2007): 71–100.

Bykov, P. V. "I. A. Salov." In I. A. Salov, *Sochineniia*, 5: 3–27. St. Petersburg: Tip. A. F. Marks, 1908.

Carr, G., ed., *New Essays in Ecofeminist Literary Criticism.* Lewisburg, PA: Bucknell University Press, 2000.

Carr, Raymond. *English Fox Hunting: A History*, rev. ed. London: Weidenfeld and Nicolson, 1986.

Carter, K. Codell. "Nineteenth-Century Treatments for Rabies as Reported in the *Lancet*." *Medical History* 26 (1982): 67–78.

Cartmill, Matt. *A View to a Death in the Morning: Hunting and Nature Through History*. Cambridge, MA: Harvard University Press, 1993.

Cavender, Mary. "Hunting in Imperial Russia: State Policy and Social Order in L. P. Sabaneev's Writing." *The Russian Review* 76 (July 2017): 484–501.

Chekhov, A. P. *Polnoe sobranie sochinenii i pisem v tridtsati tomakh*. Edited by N. F. Belchikov. Moscow, Nauka, 1974-1988.

———. "Hydrophobia." In *The Unknown Chekhov: Stories and Other Writings Hitherto Untranslated*, edited and translated by A. Yarmolinsky, 95–106. New York: Noonday, 1954.

Ch-v [Chertkov], V. "Zlaya zabava." *Novoe vremia* 5284 (November 13, 1890): 2–3.

Chernobaev, E. "S. Pospelov. Razskazy o dikikh zhivotnykh Rossii." *Vestnik Rossiiskago Obshchestva pokrovitel'stva zhivotnym* 1 (January 1906): 23-28.

Clark, Tim W., Murray B. Rutherford, and Denise Casey, eds. *Coexisting with Large Carnivores: Lessons from Greater Yellowstone*. Washington, DC: Island Press, 2005.

Clark, Timothy. *The Cambridge Introduction to Literature and the Environment*. Cambridge: Cambridge University Press, 2011.

Coleman, Jon T. *Vicious: Wolves and Men in America*. New Haven, CT: Yale University Press, 2004.

Costlow, Jane. "'For the Bear to Come to Your Threshold': Human-Bear Encounters in Late Imperial Russian Writing." In *Other Animals: Beyond the Human in Russian Culture and History*, edited by Jane Costlow and Amy Nelson, 77–94. Pittsburgh, PA: University of Pittsburgh Press, 2010.

———. "The Gallop, The Wolf, The Caress: Eros and Nature in *The Tragic Menagerie*." *The Russian Review* 56 (April 1997): 192–208.

———. *Heart-Pine Russia: Walking and Writing the Nineteenth-Century Forest*. Ithaca, NY: Cornell University Press, 2013.

Costlow, Jane and Amy Nelson, eds. *Other Animals: Beyond the Human in Russian Culture and History*. Pittsburgh, PA: University of Pittsburgh Press, 2010.

Christian, R. F. "Tolstoy and the First Step." *Scottish Slavonic Review* 20 (Spring 1993): 7–16.

Dahl, V. *Tolkovyi slovar' zhivago velikorusskago iazyka*. 4 vols. Edited by V. V. Pchelkina. Moscow: Russky iazyk, 1978-80.

Derrida, Jacques. *The Animal That Therefore I Am*. Edited by Marie-Louise Mallet and translated by David Wills. New York: Fordham University Press, 2008.

Dirke, Karin. "Where Is the Big Bad Wolf? Notes and Narratives on Wolves in Swedish Newspapers during the 18th and 19th Centuries." In *A Fairytale in Question: Historical Interactions between Humans and Wolves*, 101–18. Edited by Patrick Masius and Jana Sprenger. Cambridge, UK: The White Horse Press, 2015.

Dostoevsky, F. M. *Polnoe sobranie sochinenii v 30 tomakh*, 30 vols. Edited by V. G. Bazanov (Leningrad: Nauka, 1972-90).

Driansky, Ye. E. *Zapiski melkotravchatogo*. Edited by V. M. Kurganova. Moscow: Sovetskaya Rossiia, 1985.
Elder, John, ed. *Return of the Wolf: Reflections on the Future of Wolves in the Northeast*. Hanover, NH: Middlebury College Press, 2000.
Elick, Catherine. *Talking Animals in Children's Fiction: A Critical Study*. Jefferson, NC: McFarland & Co., 2015.
Ely, C. D. *This Meager Nature: Landscape and National Identity in Imperial Russia*. DeKalb: Northern Illinois University Press, 2003.
Emmons, Terence. "The Russian Landed Gentry and Politics." *Russian Review* 33, no. 3 (July 1974): 269–83.
Fedorov, A. E. "Materialy dlia istorii Moskovskago Obshchestva Okhoty." *Priroda i okhota*, February 1888, 1–36.
Figes, Orlando. *Natasha's Dance: A Cultural History of Russia*. New York: Picador, 2002.
Foucault, Michel. *History of Madness in the Classical Age*. Edited by Jean Khalfa. Translated by Jonathan Murphy and Jean Khalfa. London: Routledge, 2006.
Fudge, Erica. "A Left-Handed Blow: Writing the History of Animals." In *Representing Animals*, edited by Nigel Rothfels, 3–18. Bloomington: Indiana University Press, 2002.
Gardiner, Muriel, ed. *The Wolf-Man and Sigmund Freud*. London: Karnac Books, 1989.
Garrard, Greg. *Ecocriticism*. London: Routledge, 2004.
Glotfelty, C. and H. Fromm, eds. *The Ecocriticism Reader: Landmarks in Literary Ecology*. Athens: University of Georgia Press, 1996.
Gomez-Alonso, Juan. "Rabies: A Possible Explanation for the Vampire Legend." *Historical Neurology* 51 (1998): 856–59.
Gross, Aaron and Anne Vallely, eds. *Animals and the Human Imagination: A Companion to Animal Studies*. New York: Columbia, 2012.
Graves, Will N. *Wolves in Russia: Anxiety Through the Ages*. Calgary, Canada: Detselig, 2007.
Gubin, P. M. *Polnoe rukovodstvo ko psovoi okhote v trekh chastiakh*. 3 vols. Moscow: Tipo-litografiia Morits Ivanovich Neibiurger, 1890.
Guminsky, V. M. "Predislovie" to *Zapiski melkotravchatogo*, by Ye. E. Driansky, 3–20. Edited by V. M. Kurganova. Moscow: Sovetskaya Rossiia, 1985.
Helfant, Ian M. "S. T. Aksakov: The Ambivalent Proto-Ecological Consciousness of a Nineteenth-Century Russian Hunter." *Interdisciplinary Studies in Literature and Environment* 13, no. 2 (Summer 2006): 57–71.
———. *The High Stakes of Identity: Gambling in the Life and Literature of Nineteenth-Century Russia*. Evanston, IL: Northwestern University Press, 2002.
———. "That Savage Gaze: The Contested Portrayal of Wolves in Nineteenth-century Russia." In *Other Animals: Beyond the Human in Russian Culture and History*, edited by Jane Costlow and Amy Nelson, 63–76. Pittsburgh, PA: University of Pittsburgh Press, 2010.
Heller, Léonid and Anastasiia Vinogradova de La Fortelle, eds. *Zveri i ikh reprezentatsii v russkoi kul'ture: Trudy Lozannskogo simpoziuma, 2007 g.* St. Petersburg: Baltiiskie sezony, 2010.

Hodge, Thomas. "Ivan Turgenev on the Nature of Hunting." In *Words, Music, History: A Festschrift for Caryl Emerson, Part One*, edited by L. Fleishman, G. Safran, and M. Wachtel, 291–311. Stanford, CA: Stanford University Press, 2005.

Isenberg, Andrew C. "The Moral Ecology of Wildlife." In *Representing Animals*, edited by Nigel Rothfels, 48–64. Bloomington: Indiana University Press, 2002.

Ivanova, Z. N. "T. A. Kuzminskaya: Bab'ia dolia (Rasskaz krest'ianki)." In *Tolstoy-redaktor*, edited by E. Ie. Zaidenshnur, 77–87. Moscow: Kniga, 1965.

Itzkowitz, David C. *Peculiar Privilege: A Social History of English Foxhunting, 1753-1885*. Hassocks, UK: Harvester Press, 1977.

Iversen, V. *Pervoe desiatiletie rossiiskogo obshchestva pokrovitel'stva zhivotnym. Istorichesky ocherk ego deiatel'nosti v 1865-1875 gg*. St. Petersburg: Tipografiia A. M. Kotomina, 1875.

Kartashevsky, *Samaya strashnaya bolezn' (Beshenstvo). Dlia krest'ian*. Edited by Iu. I. D'iakov. Moscow: Novaya Moskva, 1926.

Kaufman, Andrew D. *Give War and Peace a Chance: Tolstoyan Wisdom for Troubled Times*. New York: Simon & Schuster, 2014.

Kh., A. "O beshenstve u liudei i zhivotnykh." *Priroda i okhota*, January 1880, 1–22.

Kishensky, D. P. "K voprosu o vodoboiazni (po arkhivnym dokumentam Moskovskoi Ekaterinskoi bol'nitsy)." *Vrach* 44 (October 29, 1887): 849–52, and 45 (November 5, 1887): 871–72.

Kornblatt, Judith Deutsch. *The Cossack Hero in Russian Literature: A Study in Cultural Mythology*. Madison: University of Wisconsin Press, 1992.

Krasnov, G. V. "Lazarevsky, Vasily Matveevich." In *Russkie pisateli 1800-1917. Biografichesky slovar'*, edited by P. A. Nikolaev, 3: 282–83. Moscow: FIANIT, 1994.

Kravchenko, Alexander V. "The Cognitive Roots of Gender in Russian." *Glossos* 3 (Spring 2002): 1–13.

Kremlev, A. N. "Sadki na rezvost' i zlobu. Doklad Pravleniiu Rossiiskago obshchestva pokrovitel'stva zhivotnym," *Vestnik Rossiiskago Obchchestva pokrovitel'stva zhivotnym*, June 1894, 170–78.

Kruuk, Hans. *Hunter and Hunted: Relationships between Carnivores and People*. Cambridge: Cambridge University Press, 2002.

Kuzminskaya, T. A. "Beshenyi volk (Istinnoe proisshestvie)." *Vestnik Evropy* 6 (June 1886): 595–612.

Lazarevsky, V. M. *Ob istreblenii volkom domashniago skota i dichi i ob istreblenii volka. Prilozhenie k Pravitel'stvennomu Vestniku*. St. Petersburg: Tipografiia Ministerstva vnutrennikh del, 1876.

LeBlanc, Ronald D. "Tolstoy's Way of No Flesh: Abstinence, Vegetarianism, and Christian Physiology." In *Food in Russian History and Culture*, edited by Musya Glants and Joyce Stetson Toomre, 81–102. Bloomington: Indiana University Press, 1997.

Leopold, Aldo. *A Sand County Almanac*. New York: Ballantine, 1966.

Linnell, John D. C. et al., eds. *The Fear of Wolves: A Review of Wolf Attacks on Humans*. Trondheim, Norway: NINA-NIKU, 2002.

Lopez, Barry Holstun. *Of Wolves and Men*. New York: Charles Scribner's Sons, 1978.

Lyon, Ted B. and Will N. Graves. *The Real Wolf: The Science, Politics, and Economics of Coexisting with Wolves in Modern Times*. Billings, MT: Ted B. Lyon [Farcountry Press], 2014.

M., O. "Zametka k volch'emu voprosu." *Priroda i okhota*, January 1880, 104–5.

Mazharov, N. V. "Po povodu pechal'nykh okhot v osen' 1879 goda," *Priroda i okhota*, January 1880, 137–41.

Machevarianov, P. M. *Zapiski psovogo okhotnika Simbirskoi gubernii*. Edited by A. V. Skomorokhova. Minsk: Polifakt, 1991.

Manning, Roberta Thompson. *The Zemstvo in Russia: An Experiment in Local Self-Government*. Cambridge: Cambridge University Press, 1982.

Maroketti, M. P. *Praktichesky i teoretichesky traktat o vodoboiazni, soderzhashchy v sebe predokhranitel'nuiu metodu ot beshenstva*. 2 vols. St. Petersburg: Tipografiia Ministerstva vnutrennykh del, 1840.

Marvin, Garry. "Cultured Killers: Creating and Representing Foxhounds." *Society & Animals* 9, no. 3 (2001): 273–92.

———. "A Passionate Pursuit: Foxhunting as Performance." *Sociological Review* 51 (2003): 46–60.

———. "Unspeakability, Inedibility, and the Structures of Pursuit in the English Foxhunt." In *Representing Animals*, edited by Nigel Rothfels, 139–58. Bloomington: Indiana University Press, 2002.

———. *Wolf*. London: Reaktion Books, 2012.

Marvin, Garry and Susan McHugh, eds. *Routledge Handbook of Human–Animal Studies*. London: Routledge, 2014.

Masius, Patrick and Jana Sprenger. "Reconstructing the Extermination of Wolves in Germany: Case Studies from Brandenburg and Rhineland-Palatinate." In *A Fairytale in Question: Historical Interactions between Humans and Wolves*, edited by Patrick Masius and Jana Sprenger, 119–40. Cambridge, UK: The White Horse Press, 2015.

———, eds. *A Fairytale in Question: Historical Interactions between Humans and Wolves*. Cambridge, UK: The White Horse Press, 2015.

Mazharov, N. V. "Proekt sadki borzykh sobak na volka." *Priroda i okhota*, February 1881, 106–12.

Mech, L. David and Luigi Boitani, eds. *Wolves: Behavior, Ecology, and Conservation*. Chicago: University of Chicago Press, 2003.

Mondry, Henrietta. *Political Animals: Representing Dogs in Modern Russian Culture*. Leiden, The Netherlands: Brill Rodopi, 2015.

Morson, Gary Saul. *Hidden in Plain View: Narrative and Creative Potentials in 'War and Peace'*. Stanford, CA: Stanford University Press, 1987.

Moskovskoe Obshchestvo Okhoty imeni Imperatora Aleksandra II. *Al'bom v pamiati piati-desiatiletniago iubileia Moskovskago Obshchestva Okhoty imeni Imperatora Aleksandra II (1862-1912)*. Compiled by A. A. Beer et al. Moscow: T-vo skoropechatni A.A. Levenson, 1913.

Murphy, Bill and Monica Wasik, *Rabid: A Cultural History of the World's Most Diabolical Virus*. New York: Viking Penguin, 2012.

Musiani, Marco, Luigi Boitani, and Paul C. Paquet, eds. *A New Era for Wolves and People: Wolf Recovery, Human Attitudes, and Policy*. Calgary, Canada: University of Calgary Press, 2009.

Nelson, Amy. "The Body of the Beast: Animal Protection and Anticruelty Legislation in Imperial Russia." In *Other Animals: Beyond the Human in Russian Culture and History*, edited by Jane Costlow and Amy Nelson, 95–112. Pittsburgh, PA: University of Pittsburgh Press, 2010.

Newlin, Thomas. "At the Bottom of the River: Forms of Ecological Consciousness in Mid-Nineteenth-Century Russian Literature." *Russian Studies in Literature* 39, no. 2 (2003): 71–90.

———.. "Swarm Life" and the Biology of *War and Peace*." *Slavic Review* 71, no. 2 (2012): 359-84.

"Original'nyi sposob istrebleniia volkov," *Lesnoi zhurnal* 5 (1877): 89.

Orwin, Donna Tussing. *Tolstoy's Art and Thought, 1847-1880*. Princeton, NJ: Princeton University Press, 2013.

Owen, Thomas C. *The Corporation Under Russian Law, 1800-1917: A Study in Tsarist Economic Policy*. Cambridge: Cambridge University Press, 1991.

Paster [Pasteur], L. *Beshenstvo*. Translated by N. B. Lanchulizdeva and M. O. Perel'man. Saratov, Russia: Saratovsky uezdnyi zemsky sanitarnyi sovet, 1891.

Pavlov, M.P. *Volk*, 2nd ed. Moscow: Agropromizdat, 1990.

Peggs, Kay. "From centre to margins and back again: critical animal studies and the reflexive human self." In Taylor, Nik and Richard Twine, eds. *The Rise of Critical Animal Studies: From the Margins to the Centre*, 36–51. London: Routledge, 2014.

Pemberton, Neil and Michael Worboys. *Rabies in Britain: Dogs, Disease and Culture 1830-2000*. Basingstoke, UK: Palgrave Macmillan, 2012.

Perkowski, Jan Louis. *Vampire Lore: From the Writings of Jan Louis Perkowski*. Bloomington: Slavica, 2006.

Pil'niak, B. "Pozemka." In *Byl'e*, 104-112. Revel', Russia: Bibliofil, 1922.

Popoff, Alexandra. *Tolstoy's False Disciple: The Untold Story of Leo Tolstoy and Vladimir Chertkov*. New York: Pegasus, 2014.

Pospelov, S. "Dva brata (Iz volch'ei zhizni). In *Razskazy o dikikh zhivotnykh. Knizhka pervaya*, 13-33, 2nd ed. St. Petersburg: Tipografiia O. N. Popovoi, 1914.

———. "Zhestokaya zabava. Travlia." *Zashchita zhivotnykh* 11 & 12 (November and December 1905): 441–53; 491–95.

"*Psovaya okhota. Sochinenie N. Reutta* [review]." *Biblioteka dlia chteniia* 76, no. 6 (1846): 24–27.

Pravilova, Ekaterina. *A Public Empire: Property and the Quest for the Common Good in Imperial Russia*. Princeton, NJ: Princeton University Press, 2014.

Quammen, David. *Monster of God: The Man-Eating Predator in the Jungles of History and the Mind*. New York: W. W. Norton, 2003.

———. *Spillover: Animal Infections and the Next Human Pandemic*. New York: Norton, 2012.

Rathgeber, W. and V. Bonvicini, compilers. *Wolk 1. Der Lasarewski-Report Zur Wolf In Rußland*. Munich: Bengelmann Verlag and W. Rathgeber, 2011.

Reutt, N. M. *Psovaya okhota*. 2 vols. St. Petersburg: Tipografiia Karla Kraiia, 1846.

Rosendale, S. *The Greening of Literary Scholarship: Literature, Theory, and the Environment*. Iowa City: University of Iowa Press, 2002.

Rosenberg, Charles E. "Framing Disease: Illness, Society, and History." In C. E. Rosenberg and Janet Golden, *Framing Disease: Studies in Cultural History*, xiii–xxvi. New Brunswick, NJ: Rutgers University Press, 1992.

Rosenholm, Arja and Sari Autio-Sarasmo, eds. *Understanding Russian Nature: Representations, Values and Concepts*. Helsinki: University of Helsinki Aleksanteri Institute, 2005.

Rothfels, Nigel, ed. *Representing Animals*. Bloomington, Indiana University Press, 2002.

"Russkie bol'nye krest'iane v Parizhe," *Moskovskie vedomosti*, March 23, 1886, 5.

Sabaneev, L. P. "Volchy vopros." *Zhurnal okhoty* 5 (1876): 42–50.

———. *Volk. Okhotnich'ia monografiia*. In his *Vse ob okhote*, edited by D. Galkina, 1340–547. Moscow: Astrel', 2011.

Salov, I. A. "Volki." In *Russky okhotnichy rasskaz*, edited by N. I. Netesnina and compiled and introduced by M. M. Odesskaya, 163–178. Moscow: Sovetskaya Rossiia, 1991.

Samoilovich, Danilo. *Nyneshny sposob lecheniia s nastavleniem kak mozhno prostomu narodu lechit'sia ot ugryzneniia beshennoi sobaki i ot uiazvleniia zmei*, 2nd ed. Moscow: Tipografiia u N. Novikova, 1783.

Scott, Craig. "The Hunt for Truth in War and Peace." *Tolstoy Studies Journal* 3 (1990): 120–23.

Semchenkov, P. "Posledny iz mogikan [Posleslovie]." In *Zapiski psovogo okhotnika Simbirskoi gubernii*, by P. M. Machevarianov, edited by A. V. Skomorokhova, 143–52. Minsk: Polifakt, 1991.

Shaw, James H. Review of *Wolves in Russia: Anxiety through the Ages*, by Will N. Graves. *The Journal of Wildlife Management* 73, no. 6 (2009): 1025–26.

Shchegolev, P. E. "Ob avtore *Zapisok Melkotravchatogo*." In Ye. E. Driansky, *Zapiski melkotravchatogo*, edited by P. E. Shchegolev, 3–35. Moscow: ZIF Zemlia i fabrika, 1930.

Sherstneva, E. V. "Pervye Pasterovskie stantsii v Rossii." *Problemy sotsial'noi gigieny, zdravookhraneniia i istorii meditsiny* 3–4 (2012): 56–59.

Shevelev, A. S. "L. Paster i smoliane (perepiska Pastera)." *Sovetskoe zdravookhranenie* 3 (1979): 56–59.

———. "O roli S. Rachinskogo v organizatsii poezdki zhitelei Smolenskoi gubernii k Pasteru." *Zhurnal mikrobiologii, epidemiologii i immunobiologii* 5 (May 1979): 113–14.

Shishkin, V. S. "Zarozhdenie, razvitie i preemstvennost' akademicheskoi zoologii v Rossii." *Zoologicheskii zhurnal* 78, no. 12 (1999): 1381–1395.

Singer, Peter. *Animal Liberation: A New Ethics for Our Treatment of Animals*. New York: Random House, 1975.

"Strashnyi sluchai," *Zhurnal okhoty* 8, no. 44 (1862): 102–4.

Taylor, Nik and Richard Twine, eds. *The Rise of Critical Animal Studies: From the Margins to the Centre*. London: Routledge, 2014.

Todes, Daniel Philip. *Pavlov's Physiology Factory: Experiment, Interpretation, Laboratory Enterprise*. Baltimore: Johns Hopkins University Press, 2002.

Tolstoy, L. N. *Polnoe sobranie sochinenii v 90 tomakh*. 90 vols. Edited by V. G. Chertkov. Moscow: Gosudarstvennoe izdatel'stvo khudozhestvennoi literatury, 1928-64.

Tolstoy, L. N. "Predislovie" [k stat'e V. Ch-va, "Zlaya zabava"]. *Novoe vremia* 5284 (November 13, 1890): 2.

Tolstoy, L. N. *War and Peace*. Edited by George Gibian. Translated by Aylmer Maude. New York: Norton, 1996.

Tolstoy, L. N. *War and Peace*. Translated by Richard Pevear and Larisa Volokhonsky. New York: Alfred A. Knopf, 2007.

Tolstoy, V. S. "Sadka na volkov 4 ianvaria 1880 g." *Priroda i khota*, February 1880, 263–66.

Turkin, N. V. "Istorichesky ocherk deiatel'nosti Imperatorskago Obshchestva razmnozheniia okhotnich'ikh i promyslovykh zhivotnykh i pravil'noi okhoty za 25-letny period sushchestvovaniia, 1872-1897." *Priroda i okhota*, January 1898, 1–115.

———. *Okhota i okhotnich'e zakonodatel'stvo v 300-letny period tsarstvovaniia doma Romanovykh*. Moscow: Imperatorskoe Obshchestvo razmnozheniia okhotnich'ikh i promyslovykh zhivotnykh i pravil'noi okhoty, 1913.

———. *Zakony ob okhote. Kriticheskoe issledovanie russkikh okhotnich'ikh zakonopolozhenii*. Moscow: Redaktsiia zhurnala *Priroda i okhota*, 1889.

———. *Zakon ob okhote 3 fevralia 1892 goda*. Moscow: Tipografiia M. G. Volchaninova, 1892.

Ventseslavsky, A. "Zaokhtenskie volki." *Zhurnal konnozavodstva i okhoty* 11 (November 1854): 95–105.

Volunin, M. "Zver', borzoi volkodav prinadlezhashchy Brigadiru Kniaziu Gavrilu Fedorovichu Boriatinskomu." *Zhurnal konnozavodstva i okhoty* 1, no. 2 (1842): 63–79.

Voronov, G. "Zversky i opasny sposob istrebleniia volkov." *Vestnik Rossiiskago Obshchestva pokrovitel'stva zhivotnym*, June 1893, 207–8.

Vul'fov, Aleksei. *Istoriia zheleznykh dorog Rossiiskoi imperii: Vse-taki stroit'!* Moscow: Ripol, 2016.

Walker, Brett L. *The Lost Wolves of Japan*. Seattle: University of Washington Press, 2005.

Weiner, Douglas. *Models of Nature: Ecology, Conservation, and Cultural Revolution in Soviet Russia*. Bloomington: Indiana University Press, 1988.

Woodward, Wendy. *The Animal Gaze: Animal Subjectivities in South African Narratives*. Johannesburg, South Africa: Wits University Press, 2008.

Wydeven, Adrian P. Review of *The Real Wolf: The Science, Politics, and Economics of Coexisting with Wolves in Modern Times*, by Ted B. Lyon and Will N. Graves. *The Journal of Wildlife Management* 80, no. 7 (2016): 1334–35.

Zaitsev, B. K. "Volki." In *Russky okhotnichy rasskaz*. Edited by N. I. Netsnina and compiled and introduced by M. M. Odesskaya, 158–62. Moscow: Sovietskaya Rossiia, 1991.

Zinovieva-Annibal, L. D. *The Tragic Menagerie*. Translated and introduced by Jane Costlow. Evanston, IL: Northwestern University Press, 1999.

Zinovieva-Annibal, L. D. *Tragichesky zverinets*. St. Petersburg: Ory, 1907.

Z-o-i, Z-o-i. "Travli zhivotnykh." *Vestnik Rossiiskago Obshchestva pokrovitel'stva zhivotnym* 2 (February 1890): 31–36.

Index

A

"About Rabies in People and Animals" (by A. Kh.), 70, 78–79
Aderkas, M. F., 103–104
  "A Wicked Pastime" (1891), 103; *see also* Chertkov, Vladimir Grigor'evich; Pospelov, S.
Alexander II (Tsar), xi, 36, 149n113
Alexander III (Tsar), 93
American predator eradication programs, xv, 64
"Animal Baiting," (1890), 110
anthropomorphism, 52, 99, 113, 115–116, 118–119, 131
anti-wolf campaigns (views), xv, 46, 48, 97, 152n199
Arendt, Nikolai Andreevich, 73, 76–77
Armstrong, Philip, xxiv, 135–136, 161n394, 162n395
  "The Gaze of Animals," xxiv, 135

B

Bakhtin, Mikhail Mikhailovich, dialogism, theory of, 131
The Beast in the Boudoir, *see* Kete, Kathleen
"Beast: The Borzoi Wolfhound of Brigadier General Prince G. F. Boriatinsky" (1842), *see* Memnon Volunin
Belov, V., 44
  "K volch'emu voprosu" (1878), 44
Berger, John,
  "Why Look at Animals?" (1977), xxiii–xxiv, 135
Bibikov, Dmitri Ivanovich, xix, 142n13, 142n16
  The Wolf: History, Systematics, Morphology, Ecology, xix
Boitani, Luigi, xix
Boriatinsky, Prince Gavrila Fedorovich, 2–3, 5, 7, 19, 25, 28, 138; *see also* Volunin, Memnon
Boriatinsky, Princess, 133
Borisov, Ivan Petrovich, 10–11

borzois, xiv–xv, xxiv, 1–33, 37, 41, 46, 49, 54–55, 66, 68, 97–98, 104, 106–109, 111–112, 114, 117, 118, 133, 137–138, 144n37, 145n41, 145n42, 145n45, 145n50, 146n52, 146n56, 146–147n67, 147n70, 148n96, 148n101, 149n120, 150n151, 151n166, 151n180, 159n318, 162n40; *see also* scent hounds

C

Canadian Slavonic Papers, xviii
"canine madness," xxiii, 70
Cartmill, Matthew, 115, 159n335
Cavender, Mary, 149n126
Chekhov, Anton, xv, 70–96, 98, 107–110, 115–116, 119–123, 130, 133–134, 137, 154n231, 155n251, 155n258, 156n266
  "Hydrophobia," xxv, 70–96, 155n251, 156n265
  "An Unpleasantness," 85, 155n255
  "Whitebrow," xxvi, 98–99, 115, 120–123, 130
  "The Wolf," 72, 79
Chertkov, Vladimir Grigor'evich, xv, 99–103, 134, 140n5, 158n305
  "A Wicked Pastime" (1890), xv, 99, 101, 103, 111, 113, 134; *see also* Aderkas, M. F.; Pospelov, S.
A Children's Reader (*Detskoe chtenie*), 120
Coleman, Jon, xx, xxii, 161n387
  Vicious: Wolves and Men in America, xx
Costlow, Jane, ix, xviii, 50, 127, 140n4, 141n8, 155n259, 160n369, 161n374, 161n381
  Heart-Pine Russia: Walking and Writing the Nineteenth-Century Forest, xviii

D

Defense of Animals, 112, 114; *see also* Russian Society for the Protection of Animals (RSPA)

Derrida, Jacques, 135
Dostoevsky, Fyodor, 137–138, 162n400
    Brothers Karamazov, 137–138
Driansky, Yegor Eduardovich, xxiv, 6–9, 15, 18, 26–28, 145n43,
    Notes of a Hunter with a Small Leash of Hounds, xxiv, 6, 8, 15, 18, 146n56

E
ecocriticism, ix–x, xii, xvii–xviii, xx, xxii, 140n6
Elick, Catherine,
    Talking Animals in Children's Fiction: A Critical Study, 131
Ely, Christopher, 145n41
Emancipation of Russia's serfs (1861), 34
English foxhunting, xiv, xxiv, 5–6, 18, 29, 145n40
European Russia, xiii, xv, xvii, 2, 5, 7, 33, 38–39, 41–42, 44, 51, 56, 60–61, 65, 93, 149n118, 153n222

F
A Fairytale in Question: Historical Interactions between Humans and Wolves (2015), xix
Faresov, A., 111
Fet, Afanasii Afanasievich, 11
folk medicine and folk beliefs concerning rabies, werewolves, etc., xi–xii, xv–xvi, xviii, xx, xxv, 71–93, 151n157, 154n236
Foucault, Michel, 72, 153n227
Frenz, Rudolf Ferdinandovich, 5f3
Fudge, Erica, xxii–xxiii

G
Grabovsky, Dr., xi, 73–74, 76, 153n228
Graves, Will N.,
    Wolves in Russia, xix, 142n20, 162n401
Gubin, Pyotr Mikhailovich, 8, 14–17, 19, 23f4,
    Complete Guide to Hunting with Borzoi Hounds, 8, 19
gun hunting, xxiv, 3, 6, 11, 36–37, 46, 49, 54–56, 64–66, 68, 97, 115, 128, 130, 137–138

H
Hodge, Thomas, 146n52
hunter's memoirs (genre), xiii, xxi, xxiv
the hunting law (February 3, 1892), xiv, xxv, 34–35, 58, 62–64, 97, 152n203, 152n206–207, 153n222; *see also* Turkin, N. V.
Hunting News (*Okhotnich'ia gazeta*), xxv, 35, 57–58, 102–103, 149n126

I
Imperial Moscow Society of Naturalists (MOIP), 38, 55, 64, 128, 148n111, 149n112, 149n124; *see also* Imperial Society for the Promotion of Game and Wildlife of Economic Significance and Proper Hunting
Imperial Society for the Promotion of Game and Wildlife of Economic Significance and Proper Hunting, xxiv–xxv, 35–36, 38–40, 43, 64, 66, 107, 149n123
Isenberg, Andrew C., 132–133
Ivanov, Vyacheslav, 127
Iversen, V., 104

J
Journal of Horse Husbandry and Hunting, 2, 7, 52
Journal of Hunting, xi, 7, 38, 43, 46
    "A Horrific Event," xi

K
Kean, Hilda,
    Animal Rights, xxii
Kete, Kathleen,
    The Beast in the Boudoir, xxii
Kishensky, Dmitry Pavlovich, 85, 156n276
Kremlev, A. N., 111
Kuzminskaya, T. A., xxv, 70–96, 138, 156n267, 156n276
    "The Rabid Wolf," xxv, 70–96

L
Lazarevsky, Vasily Matveevich, 41–44, 46–47, 52, 61, 143n21, 150n132
    On the Destruction of Domestic Livestock and Wild Game by Wolves and on the Eradication of Wolves, 41
Leopold, Aldo, 132–134
    A Sand County Almanac, 132
    "Thinking like a Mountain," 132
"The Life and Adventures of One Wolf" (by A. L.), 115–116, 119, 122, 130
lyssophobia, 72

M
Machevarianov, P. M., 1, 7–8, 12, 14, 24, 29, 145n47–48, 147n67
    Notes of a Hunter with Hounds from the Simbirsk Province, 7
Maroketti, Mikhail Petrovich,
    A Practical and Theoretical Treatise on Hydrophobia, Including a Preventive Treatment against Rabies, 76

Index | 173

Marvin, Garry, xix, 5, 29, 144n37, 145n40, 146n56, 147n83
Mech, L. David, xix
Mondry, Henrietta, xviii, 155n252, 162n400
  Political Animals: Representing Dogs in Modern Russian Culture, xviii
Moscow Gazette, 93, 153n223
Moscow Hunting Society, xxiv, 35–38

N

Nature and Hunting (Priroda i okhota), i, 35, 39, 43–46, 57–58, 62, 70, 78, 107, 115–116, 148n111, 149n123, 150n151
Nelson, Amy, ix, xviii, 158n311, 158n313
Nikolskoe-Viazemskoe, 10–11
Novoe vremia (A New Age), xv, 99, 111

O

Obolensky, Dmitry Dmitrievich, 11
Odesskaya, Margarita Moiseevna, 123
Ostrovsky, Aleksandr Nikolaevich, 6
Other Animals: Beyond the Human in Russian Culture and History, ix, xvii–xviii

P

Pasteur, Louis, x, xii, xiv–xvi, xxv, 69–74, 77–80, 85–86, 90, 92–95, 97, 138, 154n246, 155n247, 156n266, 157n282, 157n290
Pavlov, Mikhail Pavlovich, xix, 142n19, 157n290
  The Wolf, xix
Peggs, Kay, 143n27
The Petersburg Gazette, 71, 79–80
Popoff, Alexandra, 99
Pospelov, S., 112–115, 134; *see also* Aderkas, M. F.; Chertkov, Vladimir Grigor'evich
  "A Cruel Pastime: Hounding," 97, 112
  Tales of Wild Animals, 113–115
    Chernobaev, E. (author of a review of *Tales of Wild Animal*, 1906), 114–115
    "Two Brothers," 113–114
Pravilova, Ekaterina, 34, 63
Pskov method (*pskovichi*), 37, 55, 119

R

rabies (rabid wolves or other animals), xi–xvi, xix–xx, xxiii, xxv, 44, 52, 61, 69, 70–97, 103, 106, 128, 130, 133–134, 138, 142n12, 142n20, 153n221–222, 153n224, 154n231, 154n236, 154n241, 154n246, 154–155n247, 155n251–252, 156n266, 156n270, 156n276, 157n282, 157n290
Reutt, N., 7, 145n45, 148n96
  Hunting with Hounds, 7, 17–18, 21, 28–29, 145n45
Rosenberg, Charles, xxv, 72–74, 79, 81, 90, 92, 155n247
  "Framing Disease," xxv, 72
Russian Society for the Protection of Animals (RSPA), xv, xxv–xxvi, 71, 98–99, 103–107, 109–112, 114, 157n294, 158n311
Russian Thought, 72
The Russian Word, 6
Russia's Wolf Problem (volchy vopros), xiii, xv, xxiv–xxv, 34–35, 41, 48–49, 57, 62, 68, 136, 137
Rzhevsky, Aleksei Andreevich, 75

S

Sabaneev, Leonid Pavlovich, vi, xxv, 35–36, 38–39, 43, 46–58, 62, 64–66, 68, 103, 114, 136, 137, 146n52, 149n126, 150n151, 151n166, 151n179–180, 151n186
  The Wolf, 46
Salov, Il'ia Aleksandrovich, 36, 65–66, 153n212
  "Wolves," 36, 65–69
scent hounds, xiv, 4, 7, 9–11, 14, 17–20, 26, 37, 46, 51, 54, 66, 68, 117, 145n45, 146n67, 150n151
Scott, Craig, 145n42
Sergei Alexandrovich (Grand Duke), 37, 149n113
Seton, Ernest Thompson, 114–115, 132–133, 159n336
  Wild Aimals I Have Known, 115, 132, 159n336
Shishkin, V. S., 46
Slavic Review, xvii
Society for the Advancement of the Field Qualities of Hunting Dogs, 111
Society for the Eradication of Animals, 112
strychnine, poisoning with, xv, 8, 35, 42–44; *see also* Valevsky, O. E
Sverchkov, Nikolai Yegorovich, if1, 27f5

T

Tikhomirov, Dmitry Ivanovich; *see A Children's Reader (Detskoe chtenie)*,
Tolstoy, Leo, xiii, xv, xxiv, 1–32, 67, 72, 83, 87, 99–103, 133–134, 137–138, 140n5, 144n33, 146n57, 147n78, 156n267, 158n298, 158n305
  Anna Karenina, 10

The Cossacks, 13
War and Peace, xiii, xv, xxiv, 1–32, 68, 101
Tolstoy, V. S., 107–108, 110, 159n318
Turkin, N. V., xxv, 35–36, 38–41, 57–66, 68, 103, 136–137, 149n123, 149n126–127, 152n199–200, 152n206–207, 153n222
  Hunting and Hunting Legislation during the 300-Year Period of the Romanov Dynasty, 58
  The Hunting Law of 3 February 1892, 58
  Hunting Laws: A Critical Study of Russian Hunting Legislation, 33, 58

V

Valevsky, O. E., 42
Vasil'iev, M.,
  Tales of Life and Nature by Russian Writers, 120
Vilnius Medical Society, xi, 73
Vladimir Alexandrovich (Grand Duke), 5f3, 38
Volunin, Memnon [Zhikharev, Stepan Petrovich], 2–3
"Beast: The Borzoi Wolfhound of Brigadier General Prince G. F. Boriatinsky" (1842), 2, 7, 11, 25–26, 28
Voronov, G., 105–106
"A Savage and Dangerous Method of Destroying Wolves," 105–106

W

Walker, Brett,
  The Lost Wolves of Japan, xx
Wolves: Behavior, Ecology, and Conservation, xix

Z

Zaitsev, Boris Konstantinovich, 115, 123, 160n353
  "Wolves," 115, 123–127, 131, 160n356
Zinovieva-Annibal, Lidia Dmitrievna, 115, 127, 131, 134, 137
  The Tragic Menagerie, vi, 115, 127
  "Wolves," 115, 127, 131–132
zemstvo councils, 34, 42, 44–45, 61, 148n110

www.ingramcontent.com/pod-product-compliance
Lightning Source LLC
Chambersburg PA
CBHW051117230426
43667CB00014B/2622